P9-CEM-813

POWER
TO THE
PEOPLE

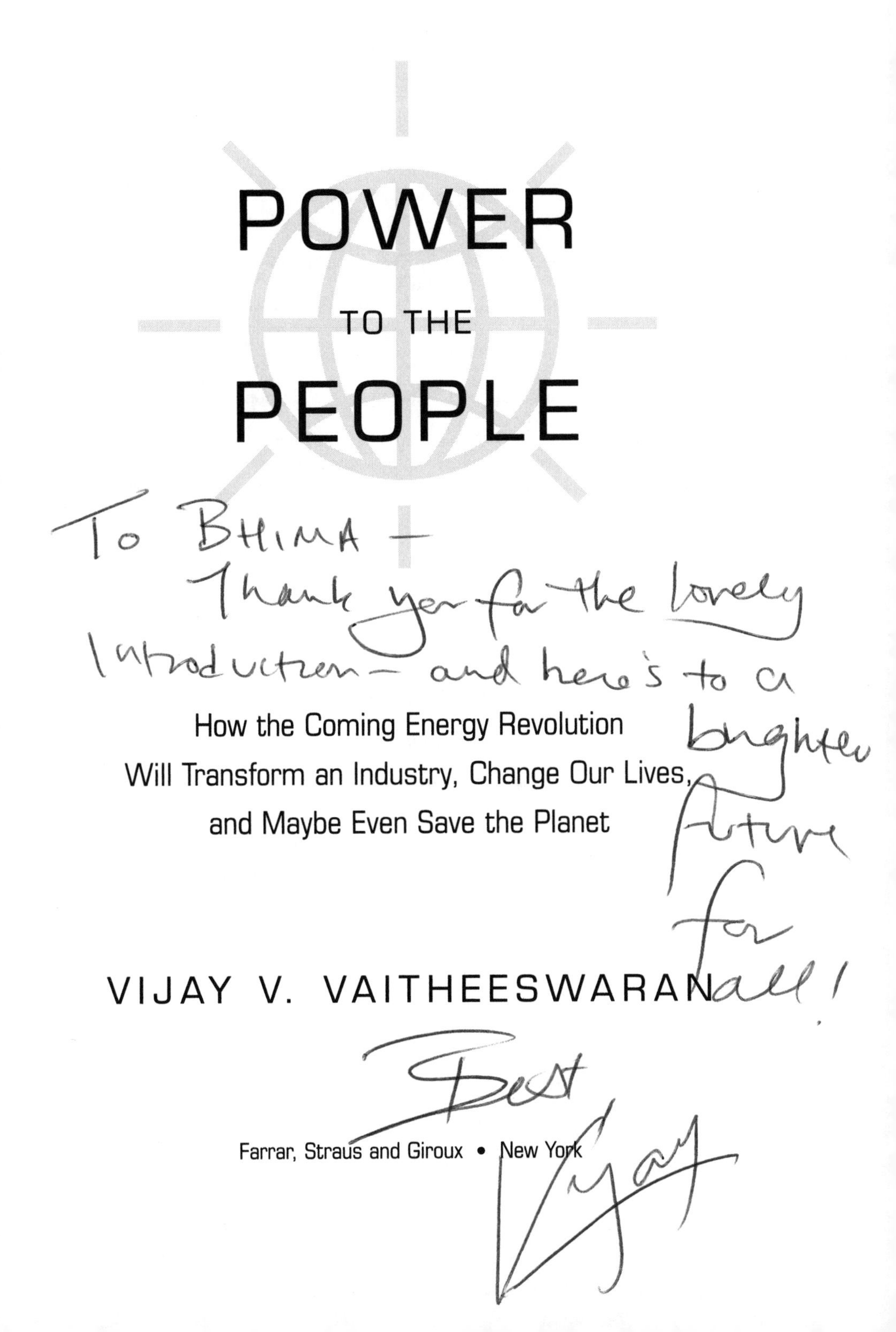

POWER TO THE PEOPLE

How the Coming Energy Revolution Will Transform an Industry, Change Our Lives, and Maybe Even Save the Planet

VIJAY V. VAITHEESWARAN

Farrar, Straus and Giroux • New York

Farrar, Straus and Giroux
19 Union Square West, New York 10003

Distributed in Canada by Douglas & McIntyre Ltd.
Printed in the United States of America
First edition, 2003

Library of Congress Cataloging-in-Publication Data
Vaitheeswaran, Vijay V., date.
Power to the people : how the coming energy revolution will transform our lives, turn the energy business upside down, and perhaps even save our planet / Vijay V. Vaitheeswaran.— 1st US ed.
p. cm.
Includes bibliographical references and index.
ISBN 0-374-23675-5
[1. Power resources. 2. Hydrogen as fuel.] I. Title.

TJ163.2.V335 2003
333.79—dc21

2003006985

Designed by Debbie Glasserman

www.fsgbooks.com

1 3 5 7 9 10 8 6 4 2

For my family

CONTENTS

POWER
TO THE
PEOPLE

Introduction: The Coming Energy Revolution

THIS BOOK IS about the future of our planet. The needlessly filthy and inefficient way we use energy is the single most destructive thing we do to the environment. Whether it is the burning of coal in industrial power plants or the felling of tropical forests, our appetite for energy—which is essential to modern life—seems insatiable. With enough clean energy, most environmental problems—not just air pollution or global warming but also chemical waste and recycling and water scarcity—can be tackled, and future economic growth can be made much more sustainable.

The problem is that change comes slowly in the energy realm. Old ways of thinking have encouraged monopolies, shielded polluters, and stifled innovation. That has burdened the rich world with an energy system locked into outmoded technologies—such as America's many coal plants—that are dirty and inefficient. That's bad enough, but now it seems that giants of the developing world, like China and India, may follow the same path as their economies surge over the next couple of decades. If they do, then many millions of unfortunates will die needlessly from the resultant pollution—as will the world's hopes of curbing the carbon emissions that are fueling global warming. That is why this is the key question: Can we move beyond today's dirty energy system to one that is cleaner, smarter, and altogether more sustainable?

Absolutely. Though cries of shortage and crisis are often heard these days in the energy world, there is actually more reason for hope than there has been in decades. This book argues that there are three powerful trends going on below the radar that promise to rewrite the rules of the energy game: the global move toward the liberalization of energy markets, the growing popular appeal of environmentalism, and the recent surge of technological innovation in areas such as hydrogen fuel cells. Taken together, they could lead to an energy system that meets the needs and desires of future generations while still tackling serious problems like global warming and local air pollution. If this clean energy revolution is really going to take off, though, we must first be ready to think the unthinkable: we must end our addiction to oil. Ironically, it may happen for reasons entirely unrelated to concerns about the environment and human health.

The problem is economic and political as much as ecological. Consider a simple question: How much is a barrel of oil worth? You might think that the price would be whatever the market will bear. Yet the price of oil is influenced less by the free interplay of supply and demand than by the whims of the Organization of Petroleum Exporting Countries (OPEC)—the ill-disciplined cartel led by Saudi Arabia. Small wonder, then, that the oil price has yo-yoed, from around $20 a barrel for much of the 1990s down to $10 in 1998 to more than $30 a barrel in early 2003.

If you could ask Osama bin Laden that same question, though, you would get a very precise figure: $144. Several years ago, before the al Qaeda terrorist group carried out its attacks on America, bin Laden made some curious comments on energy economics. In that little-noticed diatribe, he accused the United States of "the biggest theft in history" for using its military presence in Saudi Arabia to keep oil prices down. He calculated that this hostile takeover of his country's patrimony added up to some $36 trillion in lost revenues—and, he insisted, America now owes each and every Muslim in the world around $30,000. And counting.

That chilling calculation points to the nightmare scenario that keeps "energy security" experts up at night: a hostile regime seizes the oil fields of the Middle East and either raises prices sky-high or cuts off oil supplies altogether. Before September 11, scenario planners reassured themselves that if this ever happened, America would just send in its troops to quash the troublemakers and ensure safe passage for the oil supplies. After all, that was the main outcome of the Gulf War, when the coalition led by the elder George Bush booted Saddam Hussein out of the oil fields of Kuwait. And when George W. Bush began to prepare for an invasion of Iraq a decade later, even those who agreed that Saddam Hussein should be ousted took note of the fact that Iraq happens to have a bit of oil: the largest reserves in the world, in fact, after Saudi Arabia.

America's military supremacy is now unchallenged. Even so, the attacks of September 11 revealed the limits of American power in at least one realm: they have exposed the vulnerability of the global energy system to a postmodern oil shock. Today we have to consider the possibility that revolutionaries or terrorists could possess nuclear weapons—and might use them on American troops or the oil wells. Such an outcome could precipitate a global economic and political crisis of the sort never seen before. The good news is that such a scenario is extremely unlikely, even in light of recent events. The bad news is that it might still happen, and not even America's mighty military can prevent it. Even short of such an extreme outcome, though, the monopoly grip that petroleum has on the world's transport infrastructure might result in an energy crisis sometime over the next few decades.

Surprising as it may seem, the reason is not scarcity. Back in the 1970s, in the aftermath of the oil shocks of that decade, many people fretted that the energy was running out. With the arrival of the younger Bush in the White House, Americans once again heard talk of an energy crisis. Yet it's abundantly clear that there is enough oil to keep the world's motors humming for decades to come.

The real problem is not scarcity but *concentration*. The lion's share of that remaining oil—and most of the oil that is cheap to extract—lies under the desert sands of a small handful of countries in the Persian Gulf. Today, Saudi Arabia and its immediate neighbors sit atop nearly two-thirds of the world's proven oil reserves—that's right, *two-thirds*. However, those countries are not producing oil nearly as fast as they can. As the world continues to deplete expensive, non-OPEC oil in places like the deep waters of the Gulf of Mexico and the frigid reaches of Siberia in coming years, OPEC's market share is set to increase dramatically—and with it, the power of those Middle Eastern regimes. The potential for supply disruption by anti-Western terrorist bands like al Qaeda can therefore only grow. This threat is particularly acute for the United States, which is both the biggest oil guzzler and the de facto guarantor of oil supplies for its allies.

Unfortunately, there is no immediate solution, because there is no practical alternative to oil-fired transport. In the short term, all governments can do is buy some insurance against politically inspired supply disruptions and the panics that tend to accompany them. The way to do that is to expand dramatically their buffer stocks of petroleum, such as those stored in salt domes in Louisiana. To his credit, George Bush started to do this in 2001. Structural changes in the oil industry resulting from megamergers, cost-cutting, and a move to just-in-time inventories make the matter particularly urgent, because the private sector has greatly reduced its levels of stocks from the 1970s. Add to this the official neglect of government stockpiles, which are inadequate in the rich world and practically nonexistent in the developing world, and you get a world needlessly vulnerable to the next oil shock.

As for longer-term policy responses, three views typically dominate the energy debate raging around the world post–September 11: *Relax*; *Keep pumping*; and *Ride your bicycle*. The first camp insists that the very premise of the argument is false and that "energy security" is a bogus notion not worth worrying about. The second camp sees the threat as real, but argues that it can be countered effectively through supply-side measures that boost non-OPEC sources of oil. The final camp argues that conservation is the only way forward. They tend to perpetuate a number of popular myths about energy:

- The oil's about to run out
- Without fossil fuels, we'd return to the Stone Age
- Windmills and warm sweaters will save the planet
- Rampant economic growth is the root cause of our environmental problems
- Clean technologies will emerge spontaneously, without the need for government action or difficult policy measures like energy taxes
- Sport-utility vehicles (SUVs) are the work of the devil

This book will explode these and other nonsensical notions, and explain why none of these three camps gets it quite right.

What, Me Worry?

Perhaps unsurprisingly, Saudi Arabia is at the forefront of the Panglossian camp. In 1999, Ali Naimi, its oil minister, gave a speech in which he vigorously challenged the notion that his country's growing market power will be a problem: "Oil is a global market . . . those who propagate the issue of supply insecurity, dangers of import dependence and perceived instability of the Arabian Gulf are ignoring realities."

He pointed out that his country intentionally maintains a cushion of excess capacity to counter any supply disruption. It was his country's buffer, not any non-OPEC production, he noted, that came to the rescue when previous disruptions resulted from the

Iranian revolution, the Iran-Iraq war, and the Gulf War. True, but this hardly answers the question as to what will happen if his regime is toppled by a rabidly anti-Western cabal.

Lord Browne, the boss of BP, countered such fears, observing that "however fundamentalist, a regime still needs money to look after its people." Many economists agree, insisting that oil is a "fungible" commodity that is worthless unless it gets to market. In the long term, that is doubtless true. But even short-term disruptions can wreak havoc on the world economy. For example, when the Iranian revolution booted out the shah, Iran's oil exports did in fact collapse for some time, and even years later reached only two-thirds their previous level. Just imagine the chaos if willfully irrational zealots toppled the Saudi regime—and then decided to deny themselves oil revenues in order to punish the Great Satan.

Another part of the *Relax* camp of energy policy relies on free-market arguments to make its case. Libertarians argue, quite rightly, that the pain associated with previous oil shocks had more to do with foolish policy responses by Western governments meddling in the market (by imposing oil price controls, for example) than with any actual lack of supply. On one estimate, America spent some $60 billion a year during the 1990s to guard oil from the Persian Gulf, when the actual cash value of those oil imports totaled only around $10 billion a year—a mind-boggling subsidy for fossil-fuel energy. Such folk contrast this overcautious approach with America's relaxed attitude to semiconductors: these silicon sandwiches are, after all, the backbone of the digital economy and also come chiefly from just one place (in this case, Taiwan), but America's military clearly does not guard chip plants.

All that sounds quite plausible until one considers the differences between semiconductors and petroleum: the American economy can manage fine without new semiconductors for some time, but the country would grind to a halt the minute that oil dried up. Also, semiconductor plants can be built anywhere—but oil wells can go only where there is oil. The gasoline riots that brought Britain and parts of continental Europe to a standstill in late 2000 showed how quickly a modern industrial economy (even one that produces a lot of its own oil and gas, like Britain) can be crippled when its flow of oil is interrupted. That vulnerability is as good a reason as any to start weaning the world economy off petroleum.

Supply-Side Chimera

If the first camp wants you to relax, the second camp wants to get you all riled up to *Keep pumping*. To do so, these folk have tried to hijack the concerns about energy security to support domestic energy firms. Explicitly citing the need to enhance America's "energy independence," George W. Bush tried in his early days in office to push a bill through Congress that would throw open part of the Arctic National Wildlife Refuge (ANWR) to oil drilling. Environmentalists were outraged by Bush's plan because they believed that it would inevitably spoil a pristine wilderness. Yet he redoubled his efforts after September 11, arguing that the case for Alaskan oil was only strengthened. He did not even blush when critics pointed out that the pipeline through which that oil must flow is itself more insecure than oil purchased on the global market: the pipeline has been shot at, bombed, and otherwise attacked a number of times already by drunks and delusional locals. A determined band of terrorists would probably find this vital conduit, which transports over a million barrels of oil a day to the lower forty-eight states, a nearly indefensible target.

An even bigger flaw in the Alaskan proposal was that it was based on the false premise that America could ever get close to energy independence. All the oil trapped in Alaska—for that matter, in all protected lands in the country—would not provide energy independence. America consumes a quarter of the world's oil but sits atop merely 3 percent of its proven reserves.

Even assuming that oil majors invest enough money to develop new fields in non-OPEC areas like the Gulf of Mexico and Russia, the "call on OPEC" will still double over the next twenty years. In order to meet the world's unchecked thirst for oil, leading energy forecasters are hoping that Saudi Arabia and its neighbors will invest the vast sums necessary to expand output dramatically. If they do not, their output will stagnate or decline, and the consumers of the world will pay the price. But if OPEC does crank out all that extra oil, as economic self-interest would seem to dictate, consumers will still suffer. That is because the cartel's grip on the world's oil market—and therefore its ability to dictate prices—will then grow much stronger. And Russia, which has received a lot of attention of late as a potential "anti-OPEC," simply does not have enough reserves to challenge Saudi Arabia over the long haul. Alaskan oil or not, the future of the world economy will increasingly become a gamble on Middle Eastern oil. That's surely reason enough to begin the transition away from petroleum now.

CAFE Culture

"Conservation may be a sign of personal virtue, but it is not a sufficient basis, all by itself, for a sound, comprehensive energy policy." So proclaimed Vice President Dick Cheney in April 2001. The political backlash against that speech was so great that conservation is now firmly on the American political agenda. Cheney, the most forceful of those who argued that we should keep pumping, even became the poster boy for the third camp of energy thinkers: the *Ride your bicycle* gang.

At first blush, a focus on energy conservation seems an entirely good thing for America. The United States, unlike Europe, has done little to discourage the inefficient use of fossil fuels in recent years. The country imports over 11 million barrels of oil per day. America could have reduced that greatly if it had made a serious, sustained effort at curbing oil use during the last two decades.

Still, many people will always wonder how important reducing oil demand is when compared with adding supply. That is because some people's gut instinct about the nature of depletion of natural resources may be misleading. Evar Nering, a mathematician at Arizona State University, explained to readers of *The New York Times* in 2001 that the nature of exponential growth means that curbing demand is more important than adding supply: "If consumption of an energy resource is allowed to grow at a steady 5% annual rate, a full doubling of the available supply will not be as effective as reducing that growth rate by half—to 2.5%. Doubling the size of the oil reserve will add at most fourteen years to the life expectancy of the resource if we continue to use it at the currently increasing rate, no matter how large it is currently. On the other hand, halving the growth of consumption will almost double the life expectancy of the supply, no matter what it is."

Using less oil is critical, but how exactly to do that? There is actually reason to think Cheney's skepticism about conservation is justified after all (though perhaps not for the reasons he had in mind): conservation may be morally appealing to the *Ride your bicycle* camp, but it could end up being a bad thing if it merely resulted in far less mobility, trade, and other things made possible by energy that enhance human welfare. In contrast, increasing energy efficiency is a very good thing—and policies that end subsidies or other sorts of support for inefficient or dirty technologies are even better. This is particularly true given how inefficient, in energy terms, the American economy is: Europe and Japan squeeze considerably more economic output out of the energy that they use than does the United States.

One efficiency measure that is always controversial in America is the strengthening of the Corporate Average Fuel Economy (CAFE) law: raising it for cars, and closing the loophole that allows light trucks and sport-utility vehicles (SUVs) to use more gas. The automotive industry has long fought such a move, arguing that it would impose an unacceptably high cost. Yet a look at the history of CAFE suggests otherwise. The years after Jimmy Carter's presidency saw the average fuel-efficiency of America's new car fleet rise by seven miles per gallon. From 1977 to 1985, America's GDP rose by more than a quarter even as total American imports of oil fell by two-fifths; over that period, America's productivity in oil use soared. In other words, fuel-efficiency measures need not equal disaster. Even so, a far better way to encourage efficiency would be a price signal—for example, the imposition of a higher gasoline tax designed to reflect the environmental harm and energy security risks involved in using petroleum.

The car industry put on a full-court press in Washington to kill the effort to strengthen CAFE, insinuating that it would be the death of the American car industry. However, its bluff was called by a nonpartisan study done by America's National Academy of Sciences (NAS) in 2001. That analysis debunked the industry's arguments by identifying readily available technologies that could "significantly reduce fuel consumption of new cars over the next fifteen years." The experts were certain that reductions in fuel use up to 20 percent could be achieved easily.

What's more, the NAS group left the door open for even bigger reductions if radical new technologies that are now getting close to commercialization penetrate the market. Their optimism was based on the exciting new combination of hydrogen energy and fuel-cell cars, which makes it possible for the first time to contemplate a system of personal mobility that is completely free of harmful emissions and does not rely on the iron nexus of gasoline and the internal combustion engine. If that magical technology really takes off, and it will probably take a decade or more before it hits the big time, it could signal the end of the Age of Oil—and bring with it the death of OPEC, the collapse of Middle Eastern dictatorships, and a radical realignment of geopolitics. Because the hydrogen energy required to feed those fuel cells can be produced in all sorts of ways all over the world, and not just in the Middle East, this brave new energy world would not see any wars waged over energy resources and could never be held hostage by a future Osama bin Laden.

Impossible, you say? Not at all. In fact, this energy revolution is already well under way, as a trek to the mountaintop home of Amory Lovins reveals.

The Sage of Snowmass Speaks

If you want to catch a glimpse of our planet's future, visit the Rocky Mountain Institute (RMI). Nestled away in Old Snowmass, a quaint hamlet high in Colorado's snowcapped peaks, this curious think tank and "do tank" attracts visitors from all over the world who are interested in new ideas about energy and the environment. Upon arrival, visitors often find themselves on a tour whose highlights include a superefficient toilet and an indoor banana farm, "perhaps the world's highest," as one staffer boasted without hint of irony. Despite the elevation, the people who run this place do not really have their heads stuck in the clouds.

Amory Lovins is the intellectual force behind RMI. Like all visionaries, he gets things wrong, but he has also gotten some big things spectacularly right. In an article published in *Foreign Affairs* in the gloom after the first oil shock in the 1970s, he famously predicted that improvements in energy efficiency would lead to the decoupling of economic growth and energy use. At the time, most were convinced that America would continue to suck up more energy in lockstep with economic growth, and Lovins was widely ridiculed. Even America's Department of Energy had predicted that by the year 2000, oil prices would have skyrocketed to more than $150 a barrel in today's money. Though Americans will always complain about gasoline prices above a buck a gallon at the retail pump, the DOE's predictions were clearly wrong. America has learned to use energy more efficiently than it did in the 1970s—though, it must be noted, still not as efficiently as Japan or Europe—and history has vindicated Lovins.

For some years now, the Sage of Snowmass has been making another sweeping forecast for the future of energy, and again he is sounding fanciful: "This breakthrough will be like the leap from the steam engine to the diesel locomotive, from the typewriter to the laptop computer . . . it's a really disruptive technology." He gestures toward a covered object in the center of a spacious high-tech workshop where his team of engineers has been beavering away for years. With a flourish befitting a mad scientist, he unveils his creation: the Hypercar.

After nearly a decade of work, and with the support of big industrial firms from Europe, Japan, and the United States, his outfit has developed a concept car that it believes will be the clean power plant of the future: it features electric propulsion, a 100 percent composite-plastics body, highly sophisticated electronics and software, and a radically simplified and integrated design. Most important, his roomy and stylish SUV will be powered by a stack of fuel cells.

What exactly are fuel cells? According to Lovins and others, these nifty inventions are the Next Big Thing. They are essentially big batteries that produce electricity by combining hydrogen fuel and available oxygen. They do this much more efficiently than a conventional car engine that uses gasoline. They run nearly silently. Best of all, their only by-product is harmless water vapor. They are already beginning to appear in stationary applications, such as generating power for clusters of homes and factories, and are likely to appear within a few years in portable applications: laptop computers, cellular phones, even climate-controlled bodysuits for tomorrow's soldiers.

Greens, consumers, and industrialists alike should rethink their prejudices. With fuel-cell technology, even a gargantuan Ford Expedition could sip hydrogen and emit absolutely none of the usual tailpipe gases that contribute to smog and global warming or that damage human health. There's a dream that avid consumers and righteous environmentalists might share.

But Lovins has his eye on bigger game. He is convinced that consumers will be able to use the fuel cell under the hood as a "micropower" plant that can power their homes or offices. Such cars might also be used as backup generators, or while traveling in remote areas. He sees nothing preventing consumers from plugging these electric cars into a wall socket during peak hours, when the power grid is overloaded, and selling the electricity they generate back to the utilities for a profit.

In a nutshell, Lovins thinks that some version of the Hypercar will turn the modern world upside down. It is tempting to dismiss his latest forecast as hopelessly utopian. Oddly enough, though, just days before Lovins unveiled his Hypercar on the other side of the world, another wild-haired visionary, Ferdinand Panik, had introduced a similar hyper-green power plant on wheels. At that unveiling, in Berlin, there had also been talk of revolution, and even the promise of an Energy Internet: "We can use the energy unit in this car for homes or stationary power. When linked together by smart electronics, our customers can buy and trade energy freely." Panik's boss, Jürgen Schrempp, was even more effusive: "The problem of how to ensure sufficient supply of energy that is environmentally friendly is the key challenge of the future, and we see fuel cells as the solution."

Schrempp and Panik were not pundits or pie-in-the-sky dreamers: they were, respectively, the chairman and the chief fuel-cell expert at DaimlerChrysler, one of the biggest carmakers in the world. The company has already spent $1 billion to develop its "new electric car" (NECAR), and Panik expects the company to shell out another billion or so over the next decade to ensure its success. Daimler now expects to have its first commercial fuel-cell cars on the road by 2005, and mass-market volumes in about a decade.

Daimler is far from alone. Honda, Toyota, and GM also say their fuel-cell cars will be ready by then, and others claim they will follow. A number of car firms and oil companies have jointly opened up a hydrogen refueling station for their demonstration cars near California's capital of Sacramento. There is also a similar hydrogen station near Munich's airport. Daimler's top managers claim that in twenty years' time, fuel cells will power perhaps 20 percent of all new passenger vehicles, and possibly all urban buses.

What do the stodgy old utilities think of all this airy talk? Ask Kurt Yeager, the head of the Electric Power Research Institute, which is the research body of the utility business. You might expect him to be dismissive of all this talk of micropower and Energy Internets. On the contrary, he can hardly contain his excitement: "Today's technological revolution in power is the most dramatic we have seen since Edison's day, given the spread of distributed generation, transportation using electric drives, and the convergence of electricity with gas and even telecommunications. Ultimately, this coming century will be truly the century of electricity, with the microchip as the ultimate customer."

If the lines between the auto industry and the power industry really do begin to blur, the impact on the economy, on industry, and on all our lives could be dramatic indeed. Consider just one killer statistic: the power generation capacity found under the hoods of cars in Germany or America is ten times that of all of the nuclear, coal, and gas power plants combined in those countries. In other words, Ford Motor Company alone could add more juice to America's power grid than all of America's conventional power utilities put together. That is what makes this recent pronouncement from Bill Ford—Ford's chairman and the great-grandson of the company's famous founder—such a bombshell: "I believe fuel cells will finally end the 100-year reign of the internal combustion engine."

That is nothing short of an endorsement of Lovins's vision, and the epitaph for today's motorcar—the filthy but durable workhorse of the twentieth century.

The Quiet Revolution

This book is a survey of something really big going on in the energy world. The first section looks at one of the three powerful forces behind that change: the rise of market forces. From California to Cologne to Calcutta, governments are liberalizing their cosseted energy markets and throwing open their borders to trade in gas and electricity. For example, about half of America's states, led by California, have forged ahead with some form of electricity deregulation. Europe and Japan are also liberalizing their gas and power markets in fits and starts. Though there will be some bumps along the way, the resultant outpouring of entrepreneurship, financial capital, and innovation promises to transform today's energy world beyond recognition.

The second section of the book examines how the recent surge of environmentalism is now reshaping energy. Outrage over local air pollution, from California to China, is putting pressure on governments to explore clean power and transport. Equally important has been the concern over climate change, which will require mankind to make a slow but sure shift to a low-carbon energy system over the course of this century. Many countries now look unfavorably on fossil fuels, and encourage renewable energy. However, the recent move by George Bush to kill the UN's Kyoto treaty on climate change has led many environmentalists to despair that America will never do its fair share to combat global warming. Look beyond Bush's desire to please the energy business, however, and you find that his skepticism about Kyoto is shared by many others, who also worry how much fighting global warming will cost—and wonder if it is really worth doing whatever the price.

So is there no hope for meaningful action on global warming? Have Big Oil and the Bush Administration made a mockery of the efforts to green the energy industry? On the contrary. Today's debates over climate change are but a small taste of the broader environmental challenges to be faced by the world as it tries to meet its soaring energy needs, and a sign that Big Oil must change—or find itself relegated to the rubbish heap of history. The most promising development on this front is the growing popularity of market-based environmentalism, which applies commonsense tools of economics like cost-benefit analysis, emissions trading, and pollution taxes to problems like climate change. By leveling the energy playing field and using carrots as well as sticks to motivate companies, governments are much likelier to nudge the market in a greener direction.

The third section of the book describes the unprecedented wave of technological innovation now upending the energy business. The deregulation of markets, when combined with rising environmental demands, is spurring the development of such promising technologies as fuel cells and microturbines. Thanks to the rise of the Internet and sophisticated command, control, and communications software, the creaky old power grid is about to leapfrog into an intelligent network worthy of being the true backbone of the digital economy.

Just a few years ago, talk of the energy sector as exciting or innovative would have inspired loud guffaws from Wall Street: after all, utilities have long been considered so safe and stable (read: boring) that they used to be called widows' and orphans' stocks. Thanks to deregulation, the rules of the game are now changing at a dizzying pace. The stock market interest in "energy technology" stocks, which even produced an Internet-style bubble in the late 1990s, is a clear sign that the broader public is waking up to the potential of fuel cells.

The happy collision of markets, environmentalism, and innovation explains the most powerful trend of all in energy today: micropower, which puts small, clean power plants close to homes and factories. That may sound unremarkable, or even like common sense, to the reader—but in the energy business it is near heresy. It is in fact a dramatic reversal of the age-old utility practice of building giant power plants far from the end user. The most surprising aspect of the micropower revolution is that tomorrow's energy world will be based as much on silicon chips, software, and superconductors as on soot and sulfur. Dramatic advances in software and electronics offer new and more flexible ways to link parts of electricity systems together. Today's antiquated power grid, designed when power flowed from big plants to distant consumers, is being upgraded to handle tomorrow's complex, multidirectional flows (the result of micropower plants selling power into the grid as well as buying from it). It is this breakthrough that will finally make possible the intelligent homes and the Energy Internet of the squeaky-clean, not-too-distant future.

Bigger than the Internet

What is about to happen in the energy realm is every bit as dramatic as the telecommunications revolution of the past two decades, which, despite the recent rocky ride of telecom stocks, has brought the world such astonishing developments as cheap long-distance calls, cellular telephony, and the Internet. In fact, the coming energy revolution is quite possibly more important, for two reasons. One is that energy is the world's biggest industry, by far—America's electricity industry alone is bigger, in terms of revenues, than the country's long-distance telephony and cellular telephony businesses combined (that calculation does not even include Big Oil, Big Coal, or Big Anything Else). All told, the global energy game is nearly a $2 trillion–a–year business.

The second reason the energy revolution is so important is, of course, the impact our energy use has on the environment. The planet's health was the theme of the famous Earth Summit organized by the United Nations in Rio de Janeiro in 1992. The world's heads of state, along with thousands of activists, lobbyists, officials, scientists, and journalists, were there to push for their pet green causes—especially fighting global warming. After a decade of sketchy progress, the world's leaders gathered for a follow-up Earth Summit in Johannesburg, South Africa, in August 2002. Once again they sought to reconcile the demands of economic development with concerns about the environment—and once again energy-related problems such as global warming and local air pollution were at the top of their list of concerns.

This time, though, something interesting happened. After the usual squabbles—over whether to put the earth first or people first—subsided, the gathered heads of state hit upon a strategy that would do both: they agreed to help the world's poorest people gain access to modern energy in ways that are environmentally sustainable. In the next couple of decades, China and India will add thousands of new power plants and many millions of new vehicles as their economies grow. The rich world should help them do so using clean technologies like renewables and micropower. If not, a window of opportunity to set the world on a clean energy footing may be lost forever. It would kill many Indians and Chinese prematurely and needlessly, and would undermine efforts to combat global warming. It may even radically alter geopolitics if the relationship between an energy-starved China and an oil-rich Saudi Arabia begins to threaten America's web of alliances in the Middle East.

The world is at a crossroads. Decisions taken in the next few years about energy in big countries like the United States will shape the investments made in energy infrastructure around the world for a generation or more. After all, coal plants and oil refineries last for decades—and that sunk investment displaces or discourages nimbler, cleaner, and more distributed options like micropower. If we want to shift to a clean, secure, low-carbon energy system during this century, the time to start is now.

If the three camps in the energy debate remain so intransigent and shortsighted, the road ahead might prove a tortuous one. Happily, there are already signs that the dizzying pace of innovation out in the real world is bringing with it entirely new and better ways of thinking about energy that may yet render their arcane policy debates irrelevant. If micropower really takes off, then there is every reason for optimism about our planet's future. Let the revolution roll!

MARKET FORCES

The Invisible Hand Ascendant

1

Micropower—Thomas Edison's Dream Revived

IMAGINE A WORLD in which power flows not from on high, but from the masses. In such a world, important decisions would be dictated not by the whims of grandees, but by the needs and wants of ordinary people. The price of meeting those desires would be set not by bureaucrats, but by the robust interplay of supply and demand. In politics, such principles are the cornerstone of democracy. In economics, they are the foundation of capitalism. In the energy realm, however, such notions are merely the stuff of fantasy.

That is because, for the better part of a century, governments around the world have run the power business as a command-and-control monopoly along the lines of the old Soviet Union. In fact, Lenin himself once boasted that "Communism is Soviet power plus the electrification of the whole country." Even in America, the land of the free and of the free market, electricity has long been considered a "natural" monopoly; many Americans continue to receive their power not from nimble local suppliers in a competitive market, but from distant power plants and local utility monopolies. To understand why the world ended up with such a heavy-handed, centralized system of electricity provision, you need to go back in time.

One fateful day in 1884, two of the greatest inventors of any age met face to face in New York City: Nikola Tesla and Thomas Edison. Tesla was a great admirer of Edison's; indeed, Tesla had traveled all the way from Europe just to meet him, armed with a glowing letter of reference. In it, one of Edison's lieutenants in Europe gushed, "I know two great men and you are one of them; the other is this young man." The meeting should have been a pleasant one, but instead it proved the opening salvo in what would become a titanic struggle of rival electrical technologies: alternating current (AC) versus direct current (DC).

At first blush, Edison would seem to have had little to fear from the twenty-seven-year-old Serbo-Croatian immigrant. Though only a bit older, Edison was already a brilliant inventor and entrepreneur, with hundreds of patents to his name. He was successful beyond his dreams, and his Edison Electric Company was growing more powerful by the day. That was due in large part to his success in establishing many stand-alone "micropower" plants, based on his development of DC technology, in homes and offices around New York City. In a stroke of marketing genius, he built his first combined-heat-and-power plant in America just a stone's throw away from Wall Street.

The trouble was that the handsome and headstrong European had come to persuade him of the benefits of an altogether different sort of electrical technology. Tesla wanted to win Edison's backing for his vision of electrification based on multiphase alternating current. In particular, he was excited to share his concept for something that did not yet exist: a practical AC motor. The dramatic advantage promised by high-voltage AC systems was that they made transmission of large quantities of power over long distances suddenly feasible. The low voltages used by Edison's DC-only approach meant that his systems could transmit power efficiently for only about a mile—meaning each of his plants was destined to be an island unto itself.

Rather than appreciate Telsa's innovative development or find a way to co-opt it, Edison made the single greatest mistake of his life: he decided to kill it. Mistakenly believing AC power to be incompatible with his incandescent lightbulb (which he had developed in 1879) and concerned about its safety, he derailed Tesla's ambitions: "We're set up for direct current in America. People like it, and it's all I'll ever fool with . . . Spare me that nonsense. [AC is] dangerous." Though Edison gave Tesla a job that day, the two were destined to become enemies. Within a few years Edison would call on all the resources of his backers, including the financial empire of J. Pierpont Morgan himself, to prevent Tesla's technology from being developed. Edison and his British secretary, Samuel Insull, convinced New York's prison service to use AC power in the first-ever electrocution of a convict. On August 6, 1890, the murderer William Kemmler was executed in the electric chair. Linking AC with death, Edison was convinced, would prove the death of AC.

A bitter struggle broke out between rival forces that has come to be known as the Battle of the Currents. By 1888, four years after the two geniuses met, Tesla had found another financial backer. George Westinghouse, a powerful businessman from Pittsburgh who had invented the railroad air brake, was convinced of the supremacy of Tesla's vision; he struck a deal with Tesla that turned AC into a formidable force. By the turn of the century, AC had won the battle hands down. So decisive was the victory, in fact, that General Electric, the successor to the Edison Electric Company, embraced alternating current too. The technological war was over, and grid electrification based on AC technology became the backbone of the electricity network that eventually extended across the country.

Electrifying the Public

This centralized approach has had undeniable success. Most of Europe, Japan, and North America are now wired up, and near-universal access to power is taken for granted everywhere in the developed world. America's National Academy of Engineering goes so far as to proclaim grid electrification the greatest engineering achievement of the twentieth century, putting it ahead of more commonly celebrated advances like the television, the radio, the telephone, the computer, and even the Internet.

Most people don't give a second thought to the inner workings of the electricity network. Most folks probably have not pondered the topic at all, except when the occasional jolt of static electricity hits them shuffling across thick carpeting, or when they remember to jump out of the swimming pool during a lightning storm. Indeed, most people think of electricity simply as something that appears magically out of the walls of our homes and offices, at little cost and requiring no more effort than the flick of a switch.

In fact, Herculean efforts and mind-boggling investments in generation, transmission, and distribution allow us to command electricity with the authority with which the mighty Zeus himself wielded his bolts of lightning.

Any youngster might explain to you that making and distributing electricity is actually a very complicated process that takes an inevitable toll on human welfare. After all, the hugely successful animated film *Monsters, Inc.* starts with the premise that in order to produce electricity, scary (if well-intentioned) monsters frighten little children in order to harness the energy released by their anxious screams. While monsters aren't really involved in electricity production, monstrosities certainly are: nuclear stations that produce deadly radioactive waste, giant hydroelectric dams that choke the life out of rivers, and coal-fired plants that spew out all sorts of nasty pollutants that are working their way into your lungs and bloodstream right this minute. The reason most people do not think about these things when they flick a power switch is that power usually comes from many miles away. When you turn on your television or stereo, an electric current that may have had its origins in a big plant tens or even hundreds of miles away starts to flow into the device. Think of the steady stream of electrons flowing into your TV from the wall socket as akin to the steady stream of water flowing into your glass from the kitchen faucet: both have their origins far away, and both must flow through elaborate systems of interconnected "pipes" to get to your house.

The world's electricity system has evolved as a hugely centralized hub-and-spoke system, in which power is generated at distant locations and then shipped through an elaborate transmission and distribution system of heavy-duty wires, transformer stations, and so on before reaching our homes and offices. In fact, the North American electricity network, with its tens of thousands of miles of intertwined and insulated copper and steel, is one of the world's largest man-made structures. That power network and its foreign counterparts have played a crucial role in the world's economic development; indeed, such grids make our modern lifestyle possible.

Even so, the shortcomings of such a top-heavy approach to energy provision are now becoming apparent. It has largely failed developing countries, where more than a billion and a half people still lack access to grid electricity. If they are to keep darkness at bay, these unfortunates must walk miles a day to fetch wood, or they must use dirty and unhealthful fuels such as cow dung. Even urban elites in poor countries that have grid connections frequently endure power outages.

Concern over the reliability and cleanliness of central-station power generation has made the grid's limitations clear in the industrialized world as well. Lured by the promise of economies of scale, the power industry has built ever bigger power plants far from the consumers of electricity. It has too often been blind to the low efficiency rates and environmental costs of those central plants. (Many of America's giant coal plants, for example, are well over thirty years old and barely manage an efficiency rate of 30 to 40 percent; in comparison, the best combined-heat-and-electricity micropower plants can achieve double that efficiency.) The power industry has also ignored the losses dissipated as heat incurred in transporting power over wires to distant consumers, which typically amounts to more than a quarter of the cost of delivered electricity in developed countries. The government and local monopolies that have long controlled the generation, transmission, and retail distribution of power never had much incentive to encourage innovation or invest in new approaches to power delivery. Since market forces were suppressed, the gross inefficiency of energy utilities did not seem to matter terribly much.

Walt Patterson, an energy thinker at Britain's Royal Institute of International Affairs, has for many years been pointing out the frailties of this "bigger is better" approach. He explains that the rest of the world has avoided some of America's power pitfalls, but not the problem of excessive centralization: "In many ways, the U.S. was the odd man out. Its ideological fixation on buccaneering 'free enterprise' and hostility to government was almost unique. Elsewhere in the world, electricity systems frequently emerged as direct responsibilities of government—civic, municipal, and so on—or as shared responsibilities with private owners. Every local and subsequently national system was its own model, for decades. The German and Japanese systems were re-created in the 1950s as private monopolies at U.S. insistence, as part of postwar reconstruction." The common thread, he adds, is that most of the rest of the world has also set up a centralized, hub-and-spoke power system that is hostile to micropower.

The irony is that things need not have turned out this way. When Thomas Edison set up his micropower plant near Wall Street more than a century ago, he thought the best way to meet customers' needs would be with nimble, decentralized power plants in or near homes and offices. So what happened? After the tough technological battle between AC and DC came an even bloodier commercial battle. The market abuses of that swashbuckling era led to a popular backlash and sweeping reforms that were entirely justified. However, by entrenching monopolies and codifying the centralized approach to electricity provision, those new laws would needlessly snuff out micropower.

Monopoly in the Making

By the end of the nineteenth century, as rival factions led by Edison and Tesla raced breathlessly to set up operations and sign up customers, crisscrossing power lines began to spread across the eastern seaboard of the United States. Tempers flared, and dirty tricks proliferated. Things got so nasty that thugs hired by one camp would go so far as to cut down the power lines of the other. Such skulduggery was but a foreshadowing of dirtier deeds to come.

Not long before his death, Edison had said to his trusted aide Samuel Insull, "Whatever you do, Sammy, make a brilliant success of it or a brilliant failure. Just do something. Make it go." He did just that. Embracing AC technology, Insull set his sights on imitating the trick that Andrew Carnegie had earlier managed with steel: he wanted to gain control of America's electric supply system. In the process, he built a business empire that was arguably as despised as Carnegie's was in its day.

On one hand, Insull was a gifted business thinker who did much to modernize the industry's archaic and even amateurish practices. He took over Chicago Edison (which eventually became today's Commonwealth Edison), one of the numerous small regional franchises of the Edison empire. Insull quickly understood that there were too many firms scrambling around after electricity customers, and he came up with several clever innovations. First, he figured out that expanding the overall size of the pie would mean he could grab a bigger share; that led him to stimulate electricity use in various ways, including offering discounts at "off-peak" periods for new sorts of customers, such as farmers. He also recognized that he could reduce his per-unit cost of power by taking advantage of the economies of scale offered by AC motors; that led him to buy ever bigger motors from firms like General Electric and thereby undercut his rivals on price.

However, he found that he could not squeeze the maximum output and profit from his giant motors, because there were so many pesky competitors buzzing around, offering customers alternative sources of energy. If he could eliminate those rivals, he could have the whole pie for himself. He kicked off a major consolidation in the industry by gobbling up nearly two dozen utilities within a few years. He also argued publicly against competition, insisting that electricity was a "natural monopoly" in which it made no sense to have competitive suppliers: "Every home, every factory, and every transportation line will obtain its energy from one common source, for the simple reason that that will be the cheapest way to produce and distribute it."

All this led to a concentration of power in the new electric utilities that greatly worried many Americans who had recently endured much pain at the hands of unscrupulous railway operators. Progressive politicians began a campaign to rein in the new ogre of the age by cracking down on electric utilities. While some utility bosses rankled, the crafty Insull saw that embracing—and if possible co-opting—the coming tide of regulation could have benefits. This first wave of regulation actually ended up legitimizing the monopoly status of electric utilities and granting them such powers as the right to seize property (eminent domain), which had previously been a privilege of the government alone. It made electric utilities more profitable in another way too: public regulation persuaded Wall Street that these firms were less risky, so they lent them money at cheaper rates.

However, Insull overreached. Even as he expounded on the importance of regulation and consumer protection, he and his cohorts built layer upon layer of holding companies through which they furtively controlled many far-flung utilities. Since many of these firms operated across state lines, they escaped scrutiny by state regulators. Energy men like Insull tilted the playing field, so that ever greater profits flowed into ever fewer hands. By the end of the 1920s, just eight big holding companies controlled nearly three-quarters of the country's electricity supply. Insull himself was the master of manipulation: with a capital investment of under $30 million, he managed to weave together a web of shadowy holdings that gave him control of electricity assets in more than thirty states worth perhaps half a billion dollars.

In the end, Insull failed spectacularly: such was the public outrage that the federal government intervened, and he nearly ended up in jail. Taking on these holding companies became a central plank of Franklin Delano Roosevelt's candidacy for President in

1932, and he railed against "the Ishmaels and the Insulls, whose hand is against everyman's." The backlash against such excesses led to the Depression-era laws (most notably the Public Utility Holding Company Act—PUHCA) that busted up powerful utility holding companies and set in place the restrictive utility laws of the United States that lasted most of the rest of the century.

Utilities were now forced to serve all customers in their local areas, and they were forbidden from competing with each other on price, quality, or services. Prices were set by all-knowing public utility boards through "cost-plus" pricing. In other words, rather than allowing the market to set prices, these officials would ordain what the "right" price for electricity should be, based on the costs incurred by the utilities.

Politicians got the stable and secure supply that they had sought—indeed, the years after the PUHCA was put in place have been called the industry's golden years—but it ultimately proved an unwise and expensive bargain. By the 1970s, there were clear signs that the old model of regulation was not doing so well. One problem was technological stagnation, not least because the stodgy industry was not attracting the sharpest minds around. Another was its bloated cost structure, which quickly got out of control as utilities rushed to build massive nuclear power plants—the ultimate central planner's fantasy. Micropower plants can be installed quickly and cheaply, but it took well over a decade to win the necessary regulatory approvals for these nuclear Goliaths and then to build them; during this time, utilities had to pay very high rates of interest on the billions of dollars they borrowed to finance them. The result was soaring prices—average rates for power in America rose by 60 percent between 1969 and 1984, even after adjusting for inflation—and irate customers. Regulatory boards, often staffed by ill-qualified appointees and political cronies, simply could not cope with this double bind of runaway costs and a consumer backlash. The system, most interested people agreed, needed fixing.

One of the clearest voices clamoring for reform over the past couple of decades has been Thomas Casten's. For many years he headed Trigen, an independent power producer that has greatly annoyed the established players of the industry with its success in building distributed-generation plants. Along with his son Sean Casten, he wrote "Transforming Electricity"—a short and sharp manifesto for reform that uses the same title as an earlier, equally heretical book by Walt Patterson. In it, the Castens share many real-life examples that make it clear that the time has come for an energy revolution:

> To understand current power related problems, consider the following:
>
> - Cooling towers discard prodigious amounts of waste heat at a coal-fired power plant in Lemont, Illinois. Across the street, a CITGO refinery burns 8% of its crude oil feedstock to produce the heat just thrown away.
> - In 1999, U.S. industry flared enough tail gas to generate 5% to 8% of annual U.S. electric production, roughly equal to the annual electricity consumption of California.
> - Steel, glass and chemical plants burn fossil fuel and consume electricity to melt scrap iron, iron ore, sand and produce new chemicals. Recycling this heat would produce 5% of U.S. electric needs without added fuel.
> - Two refineries in Gothenburg, Sweden (Shell and Prem), provide all of the heat for over 200,000 of the 450,000 residents via a district-heating network. None of the 150 operating refineries in the U.S. recycles any significant portion of their waste heat.
>
> Taken in isolation, any one of these items could be interpreted as a minor market failure. Taken in combination, they hint at a larger problem. The authors' combined 30 years of experience has found thousands of similar anec-

> dotes throughout the energy industry. The sheer number of these incidents coupled with their assault on conventional wisdom—"can the power system be that far from optimal?"—suggests something is wrong with the current energy system worldview.

As the chorus of such voices calling for reform has gained strength, politicians around the world have been forced to respond.

Beginning in the 1980s but accelerating in the early 1990s, deregulation began to shake up America's cosseted electricity business. Pundits and populists alike began to look around at the success of deregulation in industries such as interstate trucking, natural gas, and airlines—all of which brought generally lower prices and more innovative services for customers. People were also inspired by the success of electricity liberalization abroad, especially in England, where the introduction of competition has resulted in notably lower prices and greater consumer choice without any loss of reliability so far.

The biggest force for change, however, has been the changing economics of the power industry. Amory Lovins co-authored a book called *Small Is Profitable*, which explained the technological reasons why gargantuan power plants make little financial sense these days:

> By the start of the 21st Century . . . virtually everyone in industrialized countries had electric service, and the basic assumptions underpinning the big-station logic had reversed. Central thermal power plants could no longer deliver competitively cheap and reliable electricity through the grid, because the plants had come to cost *less* than the grid and had become so reliable that nearly all power failures originated *in* the grid. Thus the grid linking central stations to remote customers had become the main driver of those customers' power costs and power-quality problems—which became more acute as digital equipment

> required extremely reliable electricity. The cheapest, most reliable power, therefore, was that which was produced at or near the customers . . .
>
> Utilities' traditional focus on a few genuine economies of scale (the bigger, the less investment per kW) overlooked larger *dis*economies of scale in the power stations, the grid, the way both are run, and the architecture of the entire system . . . The capital markets have gradually come to realize this. Central thermal power plants stopped getting more efficient in the 1960s, bigger in the '70s, cheaper in the '80s, and bought in the '90s. Smaller units offered greater economies from mass-production than big ones could gain through unit size.

So the economies of scale once achieved by building bigger and bigger power plants are now being eclipsed by other sorts of economies—through reliability and efficiency as well as mass production—offered by micropower.

The result has been a global trend to liberalize electricity markets that can be broken down into three waves of reforms: The first wave created independent power producers who were allowed to sell power to the utilities. The second wave of reform, which is now in midstream, created wholesale and retail markets for power. The third wave, just getting under way, promises to speed the arrival of smaller power plants located near the end users. Progress has come only in fits and starts, but the trend is clear: the era of monopolization, centralization, and overregulation has started to give way to market forces in electricity. That, in a nutshell, explains why micropower has once again been given the chance to blossom.

After a century that saw power stations getting ever bigger, transmission grids ranging ever wider, and the dominance of central planners growing ever stronger, a dramatic new possibility has emerged: grid-connected distributed generation, or smart micropower. In this emerging world, prices are increasingly dictated by markets, not monopolies; power is increasingly generated close to the end user and not at distant central stations; and energy is at last beginning to serve the needs of ordinary people and not just the planners.

Sound far-fetched? Consider these words penned by experts at the International Energy Agency, a respected think tank for rich countries that is better known for its conservative technical reports than its flights of fancy:

> Although they represent a small share of the electricity market, distributed-generation technologies already play a key role: for applications in which reliability is crucial, as a source of emergency capacity, and as an alternative to expansion of a local network. In some markets, they are actually displacing more costly grid electricity . . . This kind of generation has the potential to alter fundamentally the structure and organization of our electric power system.

Surprising as it may seem, micropower has passed nuclear power as the technology of choice for new power plants globally. We really could be seeing the revival of Edison's dream.

Coming Soon to Your Neighborhood . . .

What does this miraculous micropower look like in real life? A visit to the shimmering Condé Nast building in New York's Times Square will give you a hint. This newish skyscraper is best known as the headquarters of such glamorous magazines as *Vogue*, *Vanity Fair*, and *The New Yorker*. However, the sexiest thing about it is not the steady stream of supermodels and movie stars, or even its stunning private cafeteria designed by Frank Gehry. No, the most remarkable things in this place are the two big, boxy machines whirring quietly away on an unfurnished floor: fuel cells. Chapter 8 describes these wondrous machines in detail, but briefly put, they are big batteries that produce electrical energy by combining hydrogen fuel with oxygen from the air. Their great attraction is that they do so much more efficiently and cleanly than conventional power plants.

In the Condé Nast building, the two fuel cells, along with the advanced solar panels integrated cleverly into the building's façade, are living examples of micropower. Thanks to those local generation sources, the building managers have greatly reduced their dependence on conventional grid power. The architect, Robert Fox, said that he was determined to take self-empowerment even further in a forthcoming project with developer Douglas Durst. They envisioned a massive new complex, occupying an entire Manhattan block, that would be powered completely by on-site sources.

Ake Almgren favors a different flavor of micropower altogether. A soft-spoken Scandinavian, Almgren was the boss of California-based Capstone Turbine Corporation, the world's biggest manufacturer of these small generation units. His personal and professional odyssey reflects the broader trend in the energy industry—and in some ways is driving that trend. He worked for many years at ABB, a European firm best known for building power stations of the traditional gargantuan sort: hydroelectric dams, nuclear plants, and the like. Convinced that the future lay instead with small-scale generation, Almgren left a top job at ABB to head what was then an upstart firm in the nascent field of microturbines.

The clever thing about a microturbine—as opposed to the big, clunky sort of turbine that is used in traditional power stations—is that it has only one moving part, a high-speed compressor-cum-rotor that spins at up to 100,000 revolutions a minute. The simplicity of the design means that microturbines are cheap to operate and maintain—running costs can be as little as one-third of those of a comparable diesel generator. Even the problem of lubricating the one part that does move seems to have been solved. Capstone has developed a version of the device that uses sophisticated "air bearings," which require no liquid lubrication. The power crisis that plagued California in the summer of 2001 and revealed the instability of the conventional power grid provided a boost to micropower.

Capstone's ultramodern factory outside Los Angeles was overflowing with the little machines, but even Almgren admitted that the micropower trend is not just about microturbines. Investors are now pouring pots of money into Stirling engines and other small power generators, advanced flywheels and other energy-storage devices, superconducting coils and cables, and a whole host of promising developments that will boost distributed generation. In sum, the trend toward smart micropower is not the result of advances in any single technology. Rather, it is the result of an entirely different way of thinking about energy.

Show Me the Money

The above technologies are undoubtedly impressive, and the forces pushing them ahead undoubtedly real, but what hard evidence is there that micropower is really about to take off? Skepticism is certainly justified, for many readers of this book probably have never even have heard of micropower before—let alone have fuel cells tucked away in their backyard. This trend could, after all, prove ephemeral. Perhaps it is merely a hundred-year curse placed by a bitter and disillusioned Edison on his deathbed.

Maybe, but if you met men like Lord Ezra and Dan Cashdan, you'd be inclined to think again. With a zeal usually seen only among religious converts, both these men are devoted to micropower and eager to talk about it. Lord Ezra, a British politician influential in energy-policy circles, invites visitors to his posh London office (around the corner from Buckingham Palace, naturally) to introduce Micropower Ltd., a group formed with the backing of a number of European corporations interested in promoting distributed generation. Lord Ezra is convinced it is the wave of the future: "Micropower is an industry poised for takeoff. The technology is in place; entry prices are falling; the number of combined heat and power generation sites, already close to 1000 [in Britain], is growing at 9 percent a year; it is being incorporated into high-profile social housing schemes such as Kensal Green and

BedZed; and it has a strategic role to play in the U.K. energy strategy." That is a startling endorsement from a man who had once served as head of the country's monolithic Coal Board.

Dan Cashdan's transformation was equally dramatic, though the place where he professed his faith could not have been more unlike Lord Ezra's old-school offices. His declaration came at a painfully hip hotel in downtown San Francisco during a weekend when the city's joyous Gay Pride parade was spilling over into the lobby. Against that backdrop of merrymakers and pounding drum-and-bass music, he and his colleague confessed their curious affection for micropower. Both of these men gave up successful careers in other fields in order to found RealEnergy, a firm devoted to micropower. The Los Angeles–based outfit targets commercial real estate developers and makes them an offer too good to refuse: "Let us place a microgenerator in your shopping mall or office complex," said Cashdan, "and we will provide you with power that is cheaper and more reliable than grid power."

Such entrepreneurial zeal is certainly inspiring, but there is an even stronger reason to believe in micropower this time around: big money and big business are starting to flock to it. Some backers—like DTE, the parent company of the Detroit Edison utility, and Tokyo Electric Power—are giant utilities with every interest in defending the ancien régime. Others are established manufacturers of conventional power-generation equipment, such as Switzerland's ABB and America's GE. Wall Street took note of all this activity: in 2000 and 2001, "energy technology" came into its own as an investment sector as investment banks rushed to put together research teams, manage public offerings, and launch specialist investment funds. The sector even underwent its own speculative bubble before investors learned to take the long view. Despite that bump, the surge in venture-capital investment into energy technology bears a striking resemblance to the pattern seen during the early days of the telecommunications revolution that followed the breakup of the AT&T monopoly.

Walt Patterson argues that the energy industry and its regulators may finally have to rethink basic assumptions. He points out that ordinary people care about cold beer and hot showers, not electrons: "People want not merely secure supplies of fuel and electricity but secure and reliable energy services, such as comfort, illumination, motive power, refrigeration, and information handling. These energy services are not commodities purchased by the batch. They are infrastructure services, delivered by assets established and maintained for that purpose. Over time, the relevant 'competitive market' should become a market not just in short-term commodity transactions for fuels and electricity, but in longer-term contracts for energy services."

Innovative business models are beginning to emerge. Utilities that once provided only kilowatts are now offering the kind of bundled energy services that Patterson envisions. Companies that used to sell equipment, like ABB, are now considering rental and leasing options for micropower plants to remove burdens from the consumer. Forward-looking firms are already developing "microgrids" that can electronically link together dozens of micropower units, be they fuel cells or wind turbines. That will be the biggest stepping-stone to the Energy Internet of the future.

All this has spurred a virtuous circle: the arrival of serious money is fueling rapid advances in research and development, which in turn is driving down costs and bringing the sector closer to commercial viability. The costs of many forms of micro-generation have tumbled in recent years, and some now approach the industry's Holy Grail: the benchmark of under $1,000 per kilowatt achieved by conventional coal plants. As costs fall further, and current trends suggest they will, micropower will become attractive to the ordinary consumer in the developed world.

The greatest market for micropower, however, may lie in helping the developing world, where more than one and a half billion people have no access to electricity. It usually costs a fortune to build or replace electricity grids in developing countries. In such places, micropower in the form of renewable energy is already an attractive option. International agencies such as the World Bank, as well as private-sector operators and nongovernmental organizations, are devising "micro-finance" schemes to help bring electricity to the poor in such countries as Mongolia and India—a theme explored in chapter 11. The market potential for micropower is enormous, but energy reformers need to overcome some big obstacles that continue to favor the dirty old ways of producing electricity.

Time to Level the Playing Field

Reformers want governments to sweep away unfair advantages that prop up the established order through taxation, regulation, and technological standards. The development of micropower can be seriously impeded by the distorted taxation and subsidy of dirty and inefficient forms of energy. Many big coal-fired power stations around the world face no "carbon" taxes on their emissions; even worse, many of them benefit from fat subsidies. In Europe, such subsidies sometimes come directly from taxpayers; in the U.S., they come through waivers that exempt old power stations from the latest environmental regulations. Tax codes may also discriminate openly against micropower. Fuel cells, for example, attract unfavorable depreciation rates in some countries. Such perverse incentives mean that micropower has never been given a chance to establish itself.

The second snag that outrages reformers is the lack of uniform standards. One significant advantage of micropower is that it can allow generator owners to become producers as well as consumers—selling surplus electricity back to the grid when they do not need it. That requires clever electronic control systems, which now exist. But it also requires commonly accepted standards. Few countries, though, have national codes governing interconnections to their power grids. As a result, manufacturers and owners of micropower units have to deal with myriad and sometimes contradictory rules whenever they want to negotiate the right to buy and sell power. An official report from America's Department of Energy acknowledged that established incumbents can all too easily thwart newcomers by citing bogus safety concerns, imposing lengthy approval processes, or demanding outrageous fees. The DOE experts say that harmonizing standards and acting more vigorously against such anticompetitive practices is essential.

Finally, say fans of petite power plants, several important technical improvements need to be made. Today's antiquated grid—designed decades ago when power essentially flowed from big power plants to distant consumers—must be upgraded so that it can handle tomorrow's more complex, multidirectional flows. Yet in many markets, especially the United States, politicians have failed to give adequate financial incentives to grid operators to make these investments. Advances in software and electronics will play a key role in enabling micropower, offering new and more flexible ways to link different electricity systems. But first, regulators need to encourage the entrenched operators of the world's power grids to adopt modern command, control, and communications software that will facilitate the connection of any power device to the grid.

The proliferation of smart electricity meters, which measure the flow of electricity in real time and vary electricity rates accordingly, will also assist the establishment of micropower. Ahmad Faruqui, an expert in this area, argues that such meters are brilliant ideas in their own right, micropower or not: "Power costs vary with demand, and are highest on hot summer afternoons when air-conditioning peaks. Yet most consumers pay the same price during all hours of the day. If prices varied through the day, consumers would cut back during peak periods, lowering average power costs."

Some people might be wary of variable pricing for electricity, but it is not such a bizarre concept: states like Washington and Florida, as well as several European countries, have already saved a lot of money by using some form of smart metering. The practice is really not much different from the logic behind the pricing of telephone calls. Those consumers who prefer stability, such as the elderly, can opt for the fixed-price deal for a small premium. Others get a generally lower tariff—but agree to accept higher prices when costs soar during peak usage. Such a deal would give consumers a good incentive to think hard about conservation and energy efficiency. It also means that ordinary folks equipped with an Internet connection could buy and sell power, based on the arbitrage between electricity and gas prices, much as day-traders bet on shares today. More likely, their smart micropower plants, in cahoots with dozens of other friendly ones, will automatically make such choices for Jo Six-Pack as she goes about her day-to-day business.

If all such obstructions were cleared away, micropower would finally have the chance to flower. Even if that eventually happens, however, the grid would not simply disappear. Rather, the micropower renaissance (in contrast with the original vision of micropower back in Edison's day) will likely embrace, rather than displace, the existing grid. Of course, in poor countries, the grid is often so shoddy and inadequate that distributed energy could well supplant it; that would make it a true "disruptive" technology. However, in rich countries, where nearly everyone has access to electricity, micropower is much more likely to grow alongside the grid. Such a hybrid vision makes the most economic and technical sense, for it allows each means of delivering energy to reinforce the reliability and flexibility of the other. The owners of distributed generators can tap into the grid for backup power, and utilities can install micropower plants close to consumers to get around bottlenecks in their transmission and distribution networks.

A micropower revolution would not spell the immediate end of the giant power plant. Well-managed plants of all shapes and sizes will be around until the end of their useful lives. Because the existing capital stock is a sunk cost and often already paid for, the marginal costs of running existing power plants can be very low. That is why America's coal plants are likely to flourish until they collapse, or until the government closes the loopholes in its environmental laws that allow these dirty old clunkers to survive. Though

Greenpeace may protest, the future is also not entirely dim for well-run nuclear plants—a phrase that seemed an oxymoron in the past. The economics of new nuclear power plants does not make sense, but shutting down existing ones might not either. Though recent minor accidents suggest that some of today's nuclear operators have not learned from the near disaster at Three Mile Island back in 1979, most of them do have strong records of safety.

The Energy Internet Is Born

In time, though, the rise of micropower could end up changing the way electricity grids themselves operate—turning them from dictatorial monopolies into democratic marketplaces. Add a bit of information technology to a microgenerator, and it will be able both to monitor itself and to talk to other plants on the grid. Visionaries see a future in which dozens, even hundreds, of disparate micropower units are linked together in so-called microgrids (in other words, lots of Jo Six-Packs talking with each other). The technology involved is not at all pie in the sky. The Electric Power Research Institute, a research body for the power industry, is developing a microprocessor-based converter that will enable "plug and play" connection of any micropower device to the power grid, and the University of California at Irvine has already got a reliable microgrid up and running.

As energy markets liberalize, on-line energy-trading markets develop, and individual consumers win the right to select their energy suppliers, some people even see the emergence of "virtual utilities." Microgrids would allow such firms to combine the individual efficiency of micropower plants with the market power that is gained by bundling together their collective generating capacity. Whether run in competition with established utilities, or by them, such virtual utilities would, according to ABB, result in "greater system reliability, lower operating costs, reduced environmental impact, and improved overall business." Much as with the Internet, the companies that develop the technology to allow the electricity grid to perform intelligent metering and switching—and position themselves as "air-traffic controllers" for these streams of electrons—will lead the industry.

Micropower could provoke change in the energy business every bit as dramatic as the revolution that hit the world's telecommunications industry in the 1980s, after the breakup of the AT&T monopoly. (Remember the old joke stolen from Henry Ford, told during the days of Ma Bell, that you could have any color phone you want—as long as it was black?) The same forces of innovation, competition, chaos, and choice that transformed yesterday's sleepy telephone monopolies into today's high-tech predators are already apparent. Besides, electricity is every bit as important as telecommunications. America's $200 billion–plus electricity market alone, never mind other sectors of the energy business, is larger than those for cellular and long-distance telephony combined.

The analogy with telecommunications goes further. All over the world, electricity firms have invested huge sums in power stations and transmission grids, which many incumbents argue make them natural monopolies. In fact, just as the phone companies that used to make the same argument have found, their legacy may serve only to turn them into easy targets. Although power grids, like all networks, may be natural monopolies, the production of what they carry is clearly not.

The arrival of competition in telecommunications led to furious investment and innovation, helping to give rise to the digital economy. Centralized control gave way to distributed networks and technologies, which pushed power that was once jealously guarded at the center out to the periphery—even to the ordinary consumer. The same sort of centrifugal force is now taking hold of every corner of the energy industry (including even oil majors, the dinosaurs of this business, as the next chapter explains). The emerging Energy Internet will be the happy result of the collision of distributed generation with distributed networking, which has already given the world such anarchic and empowering technologies as cellular phones, personal computers, interactive media, peer-to-peer systems like Napster, and, for that matter, the Internet itself.

Thanks to deregulation, the entrenched monopolies that run large, dirty power plants are facing new competition from rivals with smaller and cleaner technologies at their command. If this encouraging trend is not somehow derailed, the result is likely to be innovative services and lower prices for consumers, and more capacity to satisfy the world's ever-growing demand for energy—some of it, ironically, caused by the spread of the Internet. And it could soon mean that power is returned, quite literally, to the people.

Take New York as a down-to-earth example. Liberalization has given customers choice of retail supplier and has boosted all sorts of innovative energy-service companies. One such firm, the 1st Rochdale Cooperative Group, aims to spread energy efficiency and clean energy to some of the 1.5 million or so New Yorkers who live in housing cooperatives and condominiums. The not-for-profit group's Green Apple Initiative hopes to install solar panels, fuel cells, and microturbines in businesses and homes all across New York City in coming years. Allen Thurgood, the group's chairman, explained his firm's aspiration: "The Green Apple initiative, which promotes demand-side energy management, sustainable energy, and distributed generation . . . will assure that customer choice empowers customer control." As energy markets around the world are liberalized, many other such creative initiatives (hard to imagine under the days of heavy-handed monopoly) are getting under way that could put clean micropower right in your neighborhood—even if you live in a concrete jungle.

Enron vs. Exxon—or, The Sleeping Giants Awaken

WHAT'S THE WORLD'S largest industry? It's not a new-economy business like information technology or telecommunications. Nor is it one of the icons of the old economy, like steel, defense, or car manufacturing. Lee Raymond, the chairman of ExxonMobil, has the answer: "Energy is the biggest business in the world. There just isn't any other industry that begins to compare."

That is no exaggeration. The World Energy Council, an umbrella body for various energy interests, estimates that global investment in energy between 1990 and 2020 will top $30 trillion. And it is not just size that distinguishes the industry, notes Raymond: "Energy is the very fuel of society, and societies without access to competitive energy suffer."

Alas, its importance has been a curse for the energy business. Until recently, governments the world over have felt that it was too "strategic" to be left to the vagaries of the market. In many ways, they have ensured that oil, gas, and electricity operated outside proper market principles. Decades of mismanagement, inefficiency, unnecessary pollution, and excessively high costs have been the result.

That sorry legacy is the reason why governments everywhere are now liberalizing energy markets, if in fits and starts. This powerful trend is encouraging consolidation, cross-fertilization, and cross-border trading on a scale the energy industry has never seen before. For all the reasons outlined in the first chapter, the resulting competition should be good for consumers, who will enjoy lower prices and more innovative services in the long run.

But what will it mean for the sleeping giants and local monopolies that have dominated the energy game for so long? The answer, you may be surprised to hear, is to be found somewhere in Texas. However, depending on which particular part of the Lone Star State you visit, you may get two radically different views of that future. The state has been home to two radically different sorts of energy firms that spent the 1990s pulling the global energy business in different directions: the brash new breed of energy traders, best exemplified by now-bankrupt Enron, and the formidable old oil majors, like Lee Raymond's Exxon.

Energy Dot.Bomb

If you thought Silicon Valley during the 1990s was a dizzyingly optimistic place, you should have visited Energy Alley. That was the name given by locals to Louisiana Avenue, a broad boulevard stretching across steamy downtown Houston. First-time visitors to the Alley often thought that Houston was on some kind of local holiday, given how few people walked the sidewalks even during weekdays at lunchtime. There were, in fact, legions of office workers scurrying to and fro—but they were hidden in the downtown area's elaborate latticework of air-conditioned skyways and underground tunnels. Things were not always quite as they seemed in Energy Alley.

Houston appeared to sprout another shiny new office building every few months during the heady late 1990s. That was thanks to the dramatic rise of energy brokers like Dynegy, Reliant, Mirant, and El Paso, all of which had offices within a few blocks of one another. The granddaddy of them all, of course, was Enron, which had not one but two skyscrapers on the Alley. Much to the annoyance of those smaller neighbors, visitors arriving at the main reception at Enron's glittering world headquarters were, for a while, met with a giant banner welcoming them to THE WORLD'S LEADING ENERGY COMPANY. Sometime in 2000 the firm's bosses decided to cast aside the fig leaf of modesty: the banner was changed to THE WORLD'S LEADING COMPANY.

That small act perfectly captured the overarching ambitions, irreverent attitude, and overflowing self-confidence of Kenneth Lay and Jeffrey Skilling, the masterminds behind Enron. It also foreshadowed the hubris that would ultimately prove the firm's undoing and lead to its bankruptcy in December 2001. To see why and how far the firm fell, however, it is important to scrutinize why and how it soared so high—and to understand the powerful wave of market liberalization that propelled the firm in the first place.

In its heyday, Enron was seen to be at the cutting edge of the energy industry. At the height of the new-economy euphoria, the top honchos from global energy companies gathered in London for a closed-door brainstorming session. Mostly the men (and they were all men) bragged, gossiped, and drank too much coffee. But when Lay rose to speak, there was absolute silence. As the boss of Enron explained his vision for how liberalization and deregulation would transform the world's energy markets, the bosses of more traditional energy companies were furiously scribbling notes. Some nodded dumbly. Others shouted out ill-formed questions. Having expanded Enron's market capitalization ninefold over the past decade, asked one boss, could he possibly top that? "We'll do it again this coming decade," Lay responded without missing a beat. Mouths fell agape.

After his talk, Kenneth Lay commented quietly to an aide, with some surprise, that "some of these guys finally seem to get it." Yes, she responded smugly, "they were even using some of our language!" If you spent enough time around top Enron brass during its glory days, you could not help but get a creepy feeling that you were in the midst of some sort of evangelical cult. And in a sense, you were. Lay, with his self-professed "passion for markets," was the cult's guru. His disciples were Enron's managers, an intelligent, extremely aggressive group of youngish professionals, all of whom "got it."

The "it" was the rise of market forces in the staid energy business. Enron was born two decades ago in Texas, birthplace of many wacky cults, following the merger of two obscure natural-gas pipeline firms. At the time, the market for gas and electricity was heavily regulated or run by government monopoly almost everywhere. But Lay, a onetime academic and government official, was convinced that things would change. So he lobbied hard for deregulation and positioned himself to capitalize on the slightest aperture or loophole in the law. Thanks to his take-no-prisoners attitude and his aggressive use of financial instruments previously unheard of in the energy industry, Enron came to dominate America's markets for wholesale gas and electricity.

The Enron strategy was as simple as it was seditious: market forces will oblige Big Oil, along with the rest of the energy industry, to split up into thousands of firms, each of which will focus on its own particular niche. Energy companies will no longer need to be capital-intensive and vertically integrated, Jeffrey Skilling loved to say, but will be "virtually integrated"—by the Enrons of the world, naturally, who will "wire those thousands of firms back together cheaply and temporarily." He summed up the vision this way: "The energy industry is on the verge of massive, massive change, and it is coming fast. We are going to hasten this fragmentation of the business, and put it back together to get lower prices for customers." Lay and Skilling sounded so much like market fanatics that it was tempting not to take them too seriously—until one considered the breathtaking heights they achieved in the 1990s by following this strategy.

In less than two decades Enron grew from irrelevance into the world's largest energy-trading company. Put simply, the firm brought together producers and consumers of energy: a power-plant operator in New England with excess capacity, say, and a manufacturing firm a thousand miles away coping with unexpectedly strong sales. In the old regulated world, these two might never have found each other; indeed, they might have been forced to deal only with a monopoly supplier or local customers. By bringing together many such buyers and sellers of energy in a vibrant marketplace, energy traders like Enron helped create a national market for energy sellers that in turn produced greater choice and lower prices for energy buyers.

By 2000 Enron commanded a 15 percent stake in gas and electricity trading on both sides of the Atlantic; no one else came close. To ensure that its aggressive traders did not miss a moment of the action, Enron thoughtfully beamed the latest news and prices to them in the elevators and in the office gymnasium (there was talk, only partly in jest, of installing catheters at each trading desk). Even after the firm sank into bankruptcy, a rival estimated that it retained a market share in gas and power trading of some 5 percent in early 2002, making it much bigger than most of its solvent rivals.

EnronOnline, the firm's Internet-based commodity-trading venture, handled several hundred million transactions in 2000, its first full year of operation. After the firm's demise, however, investigators found that some of that trading volume was the result of energy traders using such shady tactics as "round-tripping"—buying and selling the same quantity of power in order to drive up trading volumes. Even deflating the figures for such bogus trades, however, many experts and even rivals acknowledged that Enron had, against all odds, created a successful Internet platform. Indeed, given that nobody had ever traded energy products in a serious way on the Web before Enron, the firm's platform forced the energy industry's slow-moving incumbents to dance to a radically different tune.

The Dinosaur Bites Back

A rather different view of the industry's future comes from Irving, Texas, the epicenter of Exxon's $200 billion empire. Unlike Enron's publicity-hungry bosses, Exxon's boss, Lee Raymond, does not look kindly on journalists. Having secured a morning on his schedule, though, I made my way to his headquarters.

The contrast with Lay's offices could not have been greater. Enron's gaudy headquarters set amid the bustle of downtown Houston reeked of new money. As if the two eponymous skyscrapers were not enough self-promotion, the firm also paid a hefty sum for the right to put its name on Houston's major-league baseball field.

Exxon's headquarters are distinctly old school: discreet low-rise buildings tucked away in a sleepy town on perfectly manicured grounds. The cabdriver that drove me to my appointment with Raymond had dropped off visitors at the secure campus many times before, and he said the place gave him goose bumps: "It's like a country club run by the CIA."

Raymond was famous in the business for possessing the same personal traits that served Lay so well: a sharp mind, blunt talk, and, of course, overflowing self-confidence. Within moments of greeting me in his spacious executive suite, he whipped out a recent copy of *The Economist* carrying my cover story "What to Do About Global Warming." The article argued that the science of climate change, while still uncertain, was worrying enough to justify some action now; the accompanying editorial explained what would add up to a reasoned response. The coverage had criticized the Kyoto Protocol—the big UN treaty on climate change—as significantly flawed, which prompted bitter attacks the previous week from green groups. This time Raymond attacked from the other end of the ideological spectrum. He tore into the story line by line with breathtaking ferocity, berating me for everything from naïveté to reliance on faulty science to the use of misleading charts.

I mounted a vigorous defense. The others present in the room—

a senior manager and a top spokesman—were dead silent. On the critical question of the evidence for rising temperatures, I pointed to the many comprehensive, peer-reviewed studies conducted by leading experts at America's National Academy of Sciences, Britain's Meteorological Office, and the United Nations. Enraged, Raymond was shouting now at the top of his voice, "Don't you know that you just can't trust these government scientists? They all have a vested interest in perpetuating the idea of global warming!"

"So what do you suggest, Mr. Raymond," I shot back, "that I rely only on those climate scientists funded by oil companies like yours?"

The others in the room turned pale—it seemed that questioning Raymond was rare. Raymond, for his part, relaxed and even managed a smile: "Yes, I suppose if you cut out government funding and industry funding, that wouldn't leave very many climate scientists at all." The ice was broken, and we went on to have a stimulating and productive interview on the future of energy.

Raymond proved a stout defender of the energy industry's traditional business model of asset-heavy vertical integration. That meant owning lots of your own giant power plants, refineries, and other expensive and inflexible kinds of infrastructure. It also involved a top-down command-and-control approach to the business that flew in the face of the global trends toward decentralization and micropower. One of Raymond's favorite expressions sums this up: "Everyone at Exxon works for the general good—and I'm the general of that general good." Another important piece of the Exxon model, perhaps unsurprisingly, was an exclusive reliance on fossil fuels.

Raymond vigorously rejected the suggestion that Big Oil is a sunset industry run by dinosaurs. Two decades from now, he predicted, Exxon will still be in much the same shape as it is today, and will still be top dog. The company spends some $10 billion a year on proprietary technology and other investments in its vast global asset base. Not a penny of that goes to renewable energy, which Raymond described as "a complete waste of money." His firm is also unconcerned about the competition posed by fuel-cell cars powered by hydrogen (see chapter 8): his scenario planners calculated that even under the most favorable conditions, by 2020 such technology will reduce global oil consumption by less than 5 percent. He summed up: "I've been through, in my career, five 'new eras' in oil, and I guess maybe a sixth will come along. But oil and gas will continue to be the dominant energy for the next twenty-five years."

Raymond was so dogmatic in his defense of the status quo that, like the Enron evangelicals, it was tempting not to take him seriously either—until you considered the breathtaking success of his strategy over the past decade too. For years Exxon was the best-managed oil major in the world, with returns on capital employed far in excess of its rivals'. When the firm gobbled up Mobil in 1999, skeptics thought that the deal, like most mergers in most industries, would fail to deliver the promised benefits. To their surprise, ExxonMobil produced savings of more than $7 billion within a few years, far higher than it had originally predicted. In 2001 the firm made a profit of more than $15 billion—yes, profit, not revenues. That figure is one of the highest of any firm in any country since the days of the rapacious Dutch East India Company. Entrenched in its ways Exxon may be, but surely it is premature to dismiss such a powerful strategic vision out of hand.

Back to the Future

So who's right about the energy future: Enron or Exxon? Let's get real, you might be thinking right about now. Anyone who has not been hiding in a cave the last couple of years knows that Enron is gone—dead, kaput, game over—while Exxon is still holding on rock-steady near the top of the *Fortune* list of the world's biggest companies. Surely all that grandiose talk from Lay and Skilling of market revolutions is now completely discredited by the revelations of financial fraud at the top of the company, and the Enron strategy is equally bankrupt too? Surely nobody in the energy business really takes anything they said seriously? Surprise: the heart and soul of Enron's strategy (minus the fraud, of course) is alive and well in energy circles.

Take Gerard Mestrallet, the boss of the giant French multinational Suez, which has interests in water, energy, and other areas. Unlike the flash and brash Houston bosses, he is a soft-spoken, even gentle, man. While Lay preached that markets could solve all problems, Mestrallet stresses their limits: in talking about the water sector, for example, he insists that "it is absolutely irresponsible to privatize in developing countries." Indeed, when he does brag, it is not about his market capitalization but rather about his support of the International Social Charter, which strengthens unions and bans child labor. In short, he is as un-Enron as you can get. Yet ask him whether the bankrupt firm was a total sham, and you get this animated reply: "The failure of Enron did not come from energy trading!" Having studied Enron very closely during its high-flying days, his firm concluded that it had no "theoretical objection" to trading—just to the dubious business practices that Enron tacked onto it. In fact, he explained that Suez was perhaps Europe's biggest energy trader and was committed to the business despite Enron's collapse.

The reason all this might come as such news to many readers is that few headlines trumpeted what energy bosses and regulators really said about energy deregulation in the wake of Enron's collapse. The endless stories in newspapers and the talking heads on television that followed the firm's bankruptcy focused mostly on the political scandal and the accounting scandal and the pensions scandal. That was justified: the Enron collapse did indeed argue for urgent reform in campaign finance, accounting transparency, revised 401(k) pension rules, and other areas of corporate governance. However, these point to structural weaknesses in American capitalism that reach far beyond the energy sector—and cannot be carelessly pinned on energy deregulation. That is why firms in many industries, ranging from conglomerates like Tyco to communications firms like Qwest even to venerable General Electric, saw their shares plunge after Enron's collapse, as investors began to punish firms with murky accounts. After all, the bankruptcy of the Houston energy giant was quickly eclipsed as the largest ever by the demise of WorldCom, a telecommunications firm—and that had nothing whatsoever to do with energy deregulation.

The only problem with all the attention lavished on these shortcomings in Enron's accounting and political practices was that far fewer column inches were devoted to another, equally important lesson from Enron's collapse: liberalizing energy markets correctly is a sensible thing to do. You would never have guessed it if you listened to the foes of market reform, who immediately gloated that Enron's collapse was the ultimate repudiation of energy deregulation. Just days after the firm went bust, in December 2001, California's Senator Dianne Feinstein, still smarting from the fallout of her state's disastrously politicized attempt at energy deregulation (described in the next chapter), insisted that Enron's bankruptcy was "the largest indictment of electricity deregulation that I can imagine." Henry Waxman, a congressman from California, attacked a pending bill on energy deregulation—which had support from members of both political parties—as the "One Last Gift for Enron Act."

Enron was indeed the fiercest global advocate for energy deregulation, and Kenneth Lay the darling of free-marketeers everywhere (including inside the Bush White House). However, insisting that the firm's collapse proves that energy deregulation is either evil or stupid, as Feinstein and many others seemed to suggest, is wrong on two counts. First of all, it was outright fraud—not the misguided or foolish nature of its core business of energy trading—that killed Enron. Second, the firm's demise did not send other energy-trading firms running for the hills: rather, like vultures swooping in for the kill, energy-trading rivals gobbled up the market share that Enron once enjoyed. When the firm went under, EnronOnline was displaced by rivals, chief among them the IntercontinentalExchange (ICE). But Richard Spencer, ICE's chief financial officer, took the trouble to praise EnronOnline even as he buried it: "Enron validated the market, they opened the door, they let us in the door, and then they left." What is more, Spencer echoed the sentiment of the entire energy-trading industry when he insisted that the collapse of Enron did not weaken the case for energy deregulation championed so forcefully by the firm: "We hope Washington will see that it's not deregulation that's the problem," he said in early 2002. "Here the market had a failure, and it went away almost without a blip."

Of course, this certainly did not feel like a blip to the thousands of employees at Enron who lost their jobs and their pensions, or to the millions of shareholders who lost their shirts. Still, his broader point resonates: even as Enron faded from the scene, the liberalization of energy markets (and innovations like Internet-based trading) that it had long championed lived on. Later in 2002, when those other energy traders themselves suffered from the post-Enron financial crunch, big Wall Street banks, hedge funds, and long-term investors like the billionaire Warren Buffett started to enter the energy business. In other words, faith in the liberalization of energy markets remained strong, Enron or no Enron.

Consider this: the world's biggest firm in a fledgling industry collapsed almost overnight, yet the system did not go haywire. Not a single supply disruption occurred. The physical markets for energy products proved remarkably robust, and the lights stayed on. That may seem unimaginable, but that is precisely what happened in the global energy markets post-Enron.

Thanks in part to Enron's tenacious lobbying, the sweeping liberalization of energy markets on both sides of the Atlantic has greatly boosted competition in wholesale gas and power. In every bit of the business, an array of companies now fight for market share. They were able to swoop in as the industry leader faded from the scene. Joe Bob Perkins of Reliant Resources, another Houston energy firm, explained why: "The physical market for gas and power was flexible enough to hold up—even if Enron had imploded overnight—because of deregulation." Pat McMurray of the Edison Electric Institute, a lobbying group representing private-sector utilities, put it bluntly: "This is about lying, cheating, and stealing . . . it's not about the electricity market."

Revealingly, energy traders (market zealots one and all) were not the only ones who believed this; energy regulators (an altogether more circumspect lot) also agreed that deregulation was not the problem. Nora Brownell, a commissioner on America's Federal Energy Regulatory Commission, argued forcefully that Enron's collapse did not result from deregulation: "As far as I can tell, this is not about a market failure. In my mind it is a classic case of a company growing very fast and not putting in place the financial controls and management depth that was needed. In fact, the market has worked pretty efficiently." Callum McCarthy, Britain's energy regulator at the time, provided this Enron postmortem in early 2002: "We had introduced a new electricity trading system just seven months earlier, when Enron, the biggest firm on our market, collapsed. Yet the system proved flexible and robust. Enron's demise proves that liberalization works."

Perform, Die—or Clone John D. Rockefeller

Most industry leaders are placing their bets on an energy future somewhere between Enron and Exxon. They are keeping their eyes on the three variables: financial performance, convergence, and risk. Just how much the industry will change will depend on whether governments will allow these forces to take full effect, and how firms will respond to them. But judging by the impact they have had already, today's notion of an "energy company" could well be turned on its ear in coming years.

First of all, energy companies will increasingly be judged on their profits, not merely on the size of their assets or their coziness with regulators, as in the past. This applies as much to energy firms that embrace the New Economy, as Enron did with Internet-based energy trading, as it does to old-fashioned firms like Exxon. This focus on financial results explains why bosses in both the utilities and oil businesses have been going for mergers, and why they will come under increasing pressure to justify the ownership of heavy assets. These forces are already shaking up the utilities business, which in the regulated past had been the least innovative corner of the energy industry. After all, unlike oil companies (which have long been savvy global companies), most utilities in the world are local yokels. In just the past few years, Germany's Veba and Viag jumped into bed together in a $17 billion deal (and the combined entity went on to bid for Britain's Powergen), Peco Energy and Unicom merged to form America's top nuclear plant operator, Electricité de France swallowed London Electricity, and numerous Anglo-American power alliances were formed. Many more marriages seem likely as electricity and gas markets are liberalized in various parts of the world. The energy firm of the future, according to Harvey Padewer, a senior manager at Duke Energy, a big American utility, is "one that is agile, flexible, quick on its feet; one that holds assets not to milk and defend them, but only so far as they serve as a means to an end; and one that understands how to manage the risks of an increasingly commoditized business." For anyone familiar with the parochial thinking of the utility bosses of yesteryear, such a vision will come as a shock.

All very well for utilities, but is any of this talk of vigorous competition and market forces really relevant to oil companies? The world's biggest oil firms are, in fact, getting bigger: in just the past few years, Exxon swallowed Mobil in an $82 billion deal; BP paid $54 billion for Amoco and then added Arco; Total made a meal of Elf and Petrofina; Chevron took over Texaco; and Phillips and Conoco got together too.

Surely, say cynics, this is more a sign of oversized egos than of market discipline. In America, for example, many nongovernmental groups consistently opposed these recent big oil mergers. Consider this evocative argument against the ExxonMobil merger, made by Athan Manuel of the US Public Interest Research Group: "The original authors of the antitrust laws sought to avoid excessive concentrations of power. As a result, Standard Oil was broken up into 34 companies in 1911. Now, Standard Oil of New York [now known as Mobil] and Standard Oil of New Jersey [now

Exxon] are getting back together. Will the cloning and reinstallation of John D. Rockefeller as CEO be far behind?"

There is some validity to such concerns. Skeptics are also right in pointing out that oil prices are set by the OPEC cartel as much as they are by the market. Look closer, though, and you find that market forces really are beginning to change the oil industry. Oilmen are famously defensive of their fiefs and have long resisted consolidation; however, the collapse in oil prices and the rise of the institutional investor in the late 1990s made such consolidation irresistible.

The price of oil fell to nearly $10 a barrel in 1998, squeezing margins and pushing firms to slash costs. Because most big oil companies had spent much of the 1990s doing just that, they were already fairly lean. The only way to make further cost savings was through big mergers, followed by ruthless restructuring. Douglas Terreson, an equity analyst at Morgan Stanley Dean Witter, correctly forecast the breathtaking wave of mergers in an article (memorably titled "The Era of the Super-Major") he penned in early 1998: "Every decade or so, a confluence of industry dynamics conspires to produce a strategic and financial environment conducive to major consolidation activity in the integrated oil sector." He argued that the twin forces of privatization and globalization created such an environment and gave a dramatic competitive edge to a few giant global majors over the myriad smaller localized players. He pointed out that the profit per employee of the biggest oil firms—Royal Dutch/Shell, and Exxon before its merger—was nearly 50 percent higher than at smaller companies, such as Texaco, Phillips, and Chevron.

Even so, the biggest root cause of merger fever was not low oil prices, although that clearly made things more urgent. The bigger reason, Terreson noted, was that oil shares performed worse than any other group of industrial shares over most of the 1990s. In fact, until the OPEC-induced price hike sent oil-company profits soaring into the stratosphere at the very end of the decade, oil firms consistently destroyed shareholder value—that is, their returns on capital employed were lower than the cost of that capital. Institutional investors, who hold perhaps three-quarters of the shares in listed oil companies, started to respond by shifting their money into other businesses. They threatened to do so ever faster unless oil companies started offering higher returns.

When asked at a time of $11-per-barrel oil whether they would have pushed ahead with their mergers if the oil price were, say, $20 a barrel, the bosses of both Exxon and BP emphatically said yes—not just in order to cut costs, but to improve returns in other ways as well. The best evidence for that came with the takeover of Texaco by Chevron, which was announced in 2000 after oil prices had rebounded to $30 a barrel.

The Urge to Converge

As if pressure for increasing profits is not enough, bosses must also address another powerful trend transforming the energy business today: the convergence of the oil, gas, electricity, and service sectors, largely in response to the spectacular rise in demand for natural gas. Twenty years ago, Western governments mistakenly thought that gas was scarce, and decreed it too valuable to be "wasted" in power generation. No longer. Gas burns much more cleanly than oil or coal, so concerns about the environmental and health impacts of fossil fuels have boosted its use, as have recent trends in power generation. The gold standard in power generation today is set by combined-cycle turbines; tomorrow's best bet may be micropower units such as fuel cells and microturbines. All of these now rely on gas. That explains why wherever wholesale markets for gas and power have been deregulated, as in America and parts of Europe, they have converged as they have taken off.

The Financial Times conveyed a sense of the convergence mania in August 2001: "Everyone in the energy industry pokes their noses into each other's business these days . . . the long-predicted convergence of the gas and electricity parts of the energy industry has arrived. It is made inevitable by liberalization and the freedom for different parts of the energy industry to cross into each other's territory and poach each other's customers."

This was very good news for innovative energy firms, which expanded into "energy service" contracts that offer customers everything from gas to power to sophisticated hedges against inclement weather. In Europe, Electricité de France rapidly expanded its presence in gas (though only outside France, thanks to legal restrictions). Britain's Centrica, a division of the old British Gas, bundled together gas, electricity, telecommunications, and even roadside assistance. The reason for such bundling was simple: experts say that a customer who receives more than one service is often 50 to 75 percent more profitable—and less likely to switch to the competition.

Even the big oil companies, which generally shunned gas and power in the past, wanted to get in on the action. Only a few years ago, oilmen were arguing that the cultural differences between the different energy sectors were insurmountable. Yet those same oilmen started converging with gusto, in part because of relentless pressure from shareholders for financial returns.

Knowing that demand for clean-burning natural gas was likely to grow much faster than that for oil over the next couple of decades, oil bosses sought ways to increase their exposure to gas-related businesses. Upstream, firms that once used to flare off gas out in the field as a useless by-product of oil exploration sought ways to get it to market. One reason why BP gobbled up Amoco was to expand its small asset base in gas into a serious force. In power generation and marketing, Shell gained a large presence through its joint ownership (with Bechtel, a large and politically connected American construction company) of Intergen. France's TotalFinaElf bought into the power business in Argentina. Before its takeover by Chevron, Texaco had contemplated an outright merger with Duke.

Some oil majors even dabbled in retail provision of electricity. One of them was Shell. Its former chairman, Sir Mark Moody-Stuart, when asked which end of the industry's strategic divide Shell belonged to—Exxon's or Enron's—insisted that the future will see not two but three sorts of energy companies: "There will always be asset managers such as Exxon, and increasingly energy traders such as Enron," he said. However, he insisted, there will also be a hybrid third sort: "The future will also see firms with big assets and market savvy that are not wedded to either approach. Rather, they will concentrate on serving the customer in the most effective way."

Moody-Stuart thought that Shell was well placed to take the third course, which would prepare it for any longer-term shifts in the industry. That was a remarkably open stance for an oilman, especially when contrasted with Lee Raymond's stout defense of the petroleum economy. The Shell man went even further, however, and in doing so, exposed the longer-term weakness of Exxon's defensive strategy: "We want to meet our customers' needs for energy, even if that means leaving hydrocarbons behind." That forward-looking notion of seeing his business as delivering energy, rather than just petroleum, was echoed soon thereafter by BP, which launched a big advertising campaign declaring itself "Beyond Petroleum." Though oil and gas remain their bread and butter, both firms have started investing in long-term prospects like renewables and hydrogen energy.

Of course, most of their investment dollars still go into oil and gas projects. Even so, forward-looking investments from oil majors nearly the size of Raymond's, with the same powers of incumbency and sunk investments in fossil-fuel assets as Exxon, suggest a truly breathtaking vision for the future. In contrast, while Exxon can boast that it avoided Enron's fate, its defensive, status quo strategy may yet be knocked sideways by long-term changes in the energy industry. That is because of the third big force shaping the energy business today: risk.

Risky Business

Managing risk is probably the scariest task energy managers face, as the bosses of California's ailing utilities (not to mention the army of unemployed Enron traders and financial wizards) will tell you.

In the future, firms in all industries—but especially the energy industry—will live or die based on how well they manage the volatility inherent in deregulated markets. Crucially, that includes the risks involved in making the transition to such markets.

Some big energy firms already have experience in managing market risk, but many others may be overwhelmed. Chuck Watson, the chief executive of Dynegy in 2001, generously offered his services to the newcomers: "It is extremely difficult to manage the risks inherent in deregulation. You need both the expertise and the size. Because I'm trading 10 to 20 billion cubic feet of gas a day all over North America, I can manage any supply/demand dislocations much better than any single customer." Indeed, big energy firms increasingly looked to the professionals: even the giant Electricité de France turned to Louis Dreyfus, a French trading company, to help manage risks as Europe's wholesale gas and power markets slowly opened to cross-border competition. Enron's Jeffrey Skilling put it this way in 2000: "It's absolutely clear that volatility in the energy business is growing because of deregulation. It is irresponsible to shareholders not to hedge those risks." Alas, as Skilling and Watson discovered to their cost, risk can be a two-edged sword: both were forced out of their cushy jobs by angry investors.

Managing fluctuations in the price of commodities is tricky, but there is another sort of risk that even the most sophisticated energy firms may not be prepared for: the emergence of a truly disruptive innovation that changes all the rules of the game. As the experience of the past two decades in telecommunications and computing has shown, the most powerful effect of deregulating an industry can be to open the door to venture capital, nimble entrepreneurship, and technological innovation that allows the previously unimaginable to happen. Even well-run firms that dominate their industry may be caught out by disruptive technologies such as personal computers and cellular telephony, as IBM and AT&T discovered to their detriment.

If such a breakthrough happens in the energy industry, even open-minded and seemingly nimble energy giants like Shell and

BP could get caught out. "No truly disruptive innovation has ever come from the established incumbent firms in an industry—they've always been too slow to see the changes coming!" Who exactly would make such sweeping pronouncements on the future of energy? Handsome and eloquent—if slightly overenthusiastic—Jeremy Leggett was the boss of Solar Century, the leading solar energy distributor in Britain. When he had visitors at his stylish low-rise office building in south London, he insisted on taking them up to his roof—even in cold, dark, and gusty weather—to show off his shiny solar contraptions scattered hither and yon. As a businessman in an emerging sector of the energy business, he had seen at sword's length how the big boys really operate. He used to be the chief climate negotiator for Greenpeace, and throughout the 1990s he often butted heads with the energy companies as they fought the Kyoto Protocol on global warming by hook and by crook. But before that, he was an insider: trained as a geologist, he spent many years doing consulting work for the oil giants. That swirl of experiences in various corners of the industry convinced him that change could come fast—but also that many of today's winners are likely to be tomorrow's losers.

"Over there is the future," he said, pointing south to a shiny solar-powered building. "That is an Internet center for disadvantaged youth from this neighborhood that we helped set up with the Lambeth council [the local government]." Circling his arms in a figurative embrace of the various generations of solar technology on his own roof, he continued: "This is the bridge to that future." Turning around swiftly and pointing accusingly at the skyscraper by the Thames River that houses Royal Dutch/Shell's headquarters, he concluded with a flourish: "And that monstrosity with the tattered Shell flag on top . . . that is surely the past!"

Bear that tale in mind, especially the remarkable things made possible by deregulation, as the next chapter takes you deep into the heart of darkness: California's power crisis.

3

Why California Went B.A.N.A.N.A.s

"ELECTRICITY IS too essential to be left to the market. It simply is not a commodity. Do you really think markets can be trusted to deliver such a critical, life-giving resource?" When Medea Benjamin speaks, people listen. A passionate critic of California's bungled effort at electricity reform, she runs the influential activist group Global Exchange, best known for opposing globalization and the World Trade Organization. Benjamin, who ran for the U.S. Senate on the Green Party ticket with Ralph Nader, is leading a grassroots movement to put the country's entire electricity industry under municipal control.

Even compared to other controversial issues her group has tackled, her electricity campaign in 2001 hit a hot button: "Usually when we ask people in the streets to sign our petitions, they keep walking by; in this case, when they hear what it's about, they turn around and come back all fired up to sign the sheet!" As she walked down the streets of San Francisco's Mission District, she was recognized and even applauded for her energy campaign. With ballot initiatives in San Francisco, San Diego, and other cities aimed at blocking market reform and seizing local control of power utilities, her "public power" movement seemed to be the only winner from the state's horrible power crisis. Benjamin and her alliance had clearly tapped into a powerful vein of public discontent.

The first signs of a backlash against electricity deregulation had surfaced the year before from an octogenarian grandmother named Tassie Dykstra. At that time, California's emerging power catastrophe had affected only the areas in and around her hometown of San Diego. She was mad as hell, and it seemed clear that California's power brokers had better pay attention. Dykstra had lived in the state for five decades, it turned out, and had given nary a passing thought to electricity supply before the summer of 2000.

That cursed summer, electricity bills all over town shot through the roof, and the blackouts began. She explained how fellow San Diegans on a fixed income had to choose between food and electricity, and neighborhood shops started to go bankrupt. She asked a question that very few in California seemed able to answer: "Why did they mess with the electricity system in the first place?" In the months that followed, as the power crisis spread across much of the state like a virulent disease, many others began to ask the same question.

Californians were up in arms because their pathbreaking effort to deregulate their power markets had gone badly awry. None of the promised benefits of cheaper power, more reliable supply, or innovative services ever materialized in the state, but unfamiliar devils such as rate hikes and brownouts did. The state's two largest utilities racked up billions of dollars of debt after the botched deregulation. Southern California Edison was pushed to the brink of bankruptcy; the Pacific Gas and Electric company went bust altogether.

Whiz-kid programmers in Silicon Valley, a number of them from blackout-plagued India, were shocked to find their dot-com offices hit by frequent and costly power outages—sometimes without the backup generators commonplace in Bangalore. The power crisis even dragged the state's weak economy, which not long before had been the silicon-gilded envy of the world, into recession in 2001. In an ironic twist, the botched attempt at market reform actually led Governor Gray Davis, who struggled to craft a response to a crisis he inherited from his predecessor, Pete Wilson, to launch a de facto state takeover of the power business.

The situation is at once tragedy and farce, and residents outside of the Golden State should take note. More than half of America's states have followed California's lead and restructured their power sectors; others are thinking about it. Dozens of countries around the world, from Italy to Brazil to Japan, are also in the midst of such reforms. In reaction to California's problems, some neighboring states have put deregulation plans on hold. European countries have sent delegations of regulators and ministers to inspect the situation firsthand, and many have slowed liberalization out of fear that they too might catch the "California virus." Justifiably or not, California's wrenching experience has come to be seen as a litmus test for electricity deregulation.

By mid-2001, many activist groups that had previously displayed little interest in energy issues suddenly jumped on the anti-deregulation bandwagon. Columnists and commentators, most prominent among them Paul Krugman of *The New York Times*, weighed in too. In one column, Krugman made this controversial argument: "True believers insist that the power crisis of 2000–2001 . . . was not a verdict on deregulation, that it was all the fault of meddling politicians who didn't let the market work. But this claim isn't particularly convincing, mainly because it isn't true. The real lesson of the California catastrophe was that the concerns that led to regulation in the first place—monopoly power and the threat of market manipulation—are still real issues today." He concluded his column with these fighting words: "There are limits to what markets can do."

Politicians like San Francisco's mayor, Willie Brown, jumped on the anti-deregulation bandwagon and transformed overnight into advocates of Benjamin's "public power" movement. In a series of in-depth articles on California's power crisis (revealingly titled "Cautionary Tale"), *The Washington Post* captured the mood of the times: "Deregulation has become synonymous with corporate greed, government incompetence and the failure of free-market economics."

Deep in the Heart of Darkness

Determined to find out whether the enduring lesson from the Californian saga is really that electricity deregulation is completely bonkers, I went straight to the epicenter of the earthquake felt around the world: Sacramento. In contrast to Washington, D.C., where there is a physical separation of the various branches of government, in California the legislature is housed in the same capitol building as the executive. With that many politicos to deal with, it seemed only prudent to get some fortification. I swung by the neighborhood espresso joint for a stiff one, only to encounter this telling sign of the times: SORRY: WE ARE CLOSING EARLY EVERY DAY DUE TO THE ELECTRICITY CRISIS.

Despite the shortage of caffeine, the entire capitol building was abuzz with activity. On one floor, the senate's energy committee was holding emergency hearings to decide whether Enron, which played a big role in the California market before going bust, should be held in contempt for refusing to hand over secret pricing information. On another floor, the governor's staff was desperately trying to arrange a multi-billion-dollar bond to pay for the bailout of the state's power utilities. Various politicians, lobbyists, and even the governor's new special adviser on electricity, David Freeman (a wily political veteran fond of cowboy hats whom the local media immediately dubbed the "energy czar"), were all eager to make clear that they never, ever really supported deregulation in the first place. However, I needed to go underground to find the one person whose opinion I was really after: Loretta Lynch, then the head of the state's top electricity regulatory body, the Public Utilities Commission (PUC).

After much hesitation, Lynch had agreed to grant an evening interview—in the basement of the capitol building. Her view mattered because her role in this crisis, unlike that of all the other players in the game, should have been crystal-clear: she and the PUC were in charge of implementing the state's plan for market reform. Though some painted her as a mere puppet of Governor Davis and suggested that she was sabotaging market reforms through arbitrary and obstructionist rulings, I gave her the benefit of the doubt. After all, if there was one person in the state whose professional reputation and personal pride should have been most closely allied with the success of electricity reform, it was the regulator in chief. What is more, her counterparts in and outside the U.S. who had also embarked on power-sector reforms were eloquent and thoughtful advocates of liberalization. I was eager to hear her take on the fiasco, her defense of the PUC's actions, and especially her response to all of the demonization of deregulation.

In particular, I wanted her to explain to me why so many Californians failed to understand why deregulation was introduced in the first place. I also wanted to hear Lynch address Medea Benjamin's claim that electricity is too essential to be trusted to markets. That argument sounds sort of reasonable until you look around and see that there are plenty of vital resources that are subject to market forces. Food, for example, is far more necessary to life than electricity. However, every country in the world save North Korea has vigorous competition among grocery stores, open-air bazaars, street-corner bodegas, and the like.

By the time Lynch showed up in the empty cafeteria in the basement of the capitol building, it was already quite late. We exchanged pleasantries before getting to the difficult topic at hand. I asked her what really went wrong in her state's deregulation effort. After much beating about the bush, the state's top electricity regulator came clean: "It's clear to me now that we simply rushed into deregulation with a naïve faith in the market akin to the blind faith of Marxism." She insisted that scrapping deregulation altogether, and returning to the earlier regime of heavy regulation, was the only solution to what ails her state. Seeing my astonishment, she then topped herself: "I'm from Missouri, where our motto is 'Show me,' and I say show me! Where has deregulation ever worked?"

Before I could point to any success stories or grill her further on her extraordinary assertions, an aide suddenly appeared out of the shadows to inform her of an urgent meeting with legislators. She brought the interview to an abrupt close and vanished down the corridors of power.

I sat a long while in that dark cafeteria, reflecting on the significance of her words. Even the person legally charged with carrying out California's electricity reforms had come to see deregulation as a dirty word. It finally struck me that in her short outburst, she had managed to reveal three fallacies that are behind the backlash against power deregulation.

The first is a muddled attack on deregulation in general that simply does not square with the documented achievements of market reforms in various countries over the last two decades. In industries ranging from interstate trucking to natural-gas distribution to telecommunications, market reform has generally been acknowledged to be a success.

A second mistake is to think that California was the first place to try electricity deregulation: since the trailblazer failed, goes the navel-gazing argument, clearly markets and electricity do not mix. A number of countries around the world embraced market reforms in the energy sector much earlier than California—and are success stories. So too are other big American states, such as Texas and Pennsylvania, that pursued electricity deregulation with less fanfare but more intelligent frameworks.

The International Energy Agency (IEA) has produced an entire book on the topic of electricity reform in various parts of the Organization for Economic Cooperation and Development (OECD)—a club of the world's richest countries. That simple but powerful policy tome, *Competition in Electricity Markets*, starts with these words: "Virtually all OECD countries have decided to open up their electricity markets, at least to their big industrial users. In many countries electricity markets will be open to all users, including households. This is already the case in Finland, Germany, New Zealand, Norway, Sweden, England and Wales in the UK, and several states in the US and Australia. By the year 2006, more than 500 million people (and all large industrial users) in the OECD area will be entitled to choose their electricity supplier. This accounts for nearly 50% of the population of the OECD countries." It seemed that Loretta Lynch had not read that book.

How have those other reforms fared? The IEA experts conclude that while "the most significant impacts of reform are only expected to emerge in the long-term as a result of better investment decisions . . . in a short-term perspective, reforms have generally delivered their expected benefits." In Scandinavia and Britain, where reforms have had more than a decade to mature, consumers are already enjoying lower prices, greater efficiency, and more innovative services. In Britain, for example, deregulation has progressed so far that perhaps a third of households have switched suppliers to take advantage of some better offer. Across America, the figure is closer to nil.

Britain's liberal daily *The Guardian* has often attacked the country's market reforms. One recent article, however, gushed about how consumers could even choose to buy electricity from suppliers that use renewable energy: "Now that green utility suppliers are dropping their prices, even the laziest, meanest would-be eco-warrior can save the planet . . . If you can be bothered to pick up a pen to sign a bank mandate form, you, too, can join the green revolution." A report published by America's National Renewable Energy Laboratory in 2001 confirmed that American consumers could also enjoy such green choices—but only if deregulation took off at the retail level. Yet, like Loretta Lynch, *The Guardian* remained unfairly dismissive: nowhere in the article was credit given to the deregulation that makes this and other sorts of retail choice possible in the first place.

The final fallacy in Lynch's monologue was perhaps the most disheartening to hear. As many pundits do, she invoked California's experience as the clinching evidence against deregulation. Even if you argue that market forces and electricity should never go together (though the evidence presented in this chapter will suggest otherwise), you certainly cannot point to California's energy crisis as evidence. That is because of a dirty little secret that Lynch won't tell you: California never really deregulated its electricity sector.

Attack of the Killer Lobbyists

If it was not death by deregulation, how exactly did California—one of the wealthiest regions in the world—end up in an energy crisis of third-world dimensions? It is tempting to finger such men as Stephen Baum as culprits. At first blush, Baum seems to be the sort of fat cat that nongovernmental organizations of all flavors love to blame for the power crisis. He runs Sempra Energy, which controls San Diego Gas & Electric, a big utility that survived the power crisis in good financial shape—in part by raising the rates paid by poor Tassie Dykstra and her neighbors in San Diego during the early days of this crisis (though the firm later offered refunds).

While Baum may have more expensive tastes than some of his opponents, he is no more culpable for the electricity mess than they are. That is because the greed of utilities was not the only—or even the main—reason why California's reform effort went astray. There is plenty of blame to go around, as it turns out, thanks to the hyper-populist politics of California.

It is true that California's politicians were pushed into deregulation by a powerful lobbying group—but that group was not the utilities. In fact, utility bosses vigorously resisted deregulation, which they saw as a threat to their cozy operations. Under the incentives offered by the old system of regulation (the one that got the PUC's Lynch so misty-eyed back in the state capitol), the utility monopolies had every incentive to build needless power plants as expensively as possible. Cost overruns had produced such a bloated, inefficient, and expensive power industry that large commercial and industrial users of electricity finally got fed up. Outraged at having to pay some 50 percent more for their electricity than rivals in neighboring states, they pushed for deregulation. They twisted arms, they padded campaign coffers, they yelled and shouted—and they threatened to leave the state unless it did something.

In mid-2002, several experts affiliated with the University of California's Energy Institute produced an unusually apolitical history of the state's power crisis. In it, Carl Blumstein and his colleagues answer Tassie Dykstra's question of why officials wanted to "mess with the electricity system" in the first place:

> From the perspective of 1992, conditions encouraged a consideration of electricity restructuring. Rates were too high, and the idea that the entire industry had to be vertically integrated was demonstrably false. Other sectors of the economy, like trucking and telecommunications, appeared to be benefiting from less reliance on traditional regulation in favor of more reliance on market forces. With the economy in recession, and the state looking for opportunities to bolster its competitive climate and attract new industry and jobs, it seemed eminently sensible to at least consider the idea of electricity restructuring at this time.

By 1996, then governor Pete Wilson had staked his political credibility on passing a power deregulation bill.

The notion met strong resistance from many quarters, not just the utilities. However, as the governor's political hand grew stronger, most of the big lobbies decided to jump on the deregulation bandwagon before it left the station. That way, they reasoned, they could at least get some provisions added to the bill that were attractive to their particular constituencies. This process of horse trading came to a head in what the local press, mindful of the resultant disaster, later dubbed the Peace Death March.

Steve Peace is a California legislator best known as the co-writer and co-producer of the cult-classic movie *Attack of the Killer Tomatoes*. Peace was not a free-market ideologue; nor was he Pete Wilson's handmaiden. However, he was head of the chief committee involved with energy, and he made a crude political calculation: California was going to deregulate sooner or later. Tired of all the backroom politicking and bilateral deals typical of the legislative process, he decided to try an audaciously different approach: marathon public negotiations on just one mega-bill. He brought together every big group with an interest in the matter—the utilities, the environmentalists, the community activists, and so on—and in effect locked them in a room till they reached a deal.

It was not pretty, but it worked: the outcome was AB 1890, a now infamous bill that laid the groundwork for the state's power deregulation. It passed unanimously, a sure sign of successful political compromise. Yet that very success sowed the seeds of failure, for the bill tied the hands of reformers by adding all sorts of bells and whistles that were completely inconsistent with market reform. It also explains why California bungled things so badly.

Bizarre as it may seem, it was the success overseas in opening up the power sector in the early 1990s that led California into the brave new world of liberalized electricity markets. In particular, explained the utility boss Stephen Baum, it was Britain that most inspired Californian reformers: "California embraced competition as a religion and the English model as our guide."

Half British, Half-baked

Why imitate Britain? The economic liberalization begun under Margaret Thatcher made that country a role model for market reformers everywhere. In electricity, at least, that admiration was perfectly understandable. In the decade after its reforms began, Britain's wholesale power costs plunged by a third, retail electricity rates dropped significantly in real terms, and reliability did not suffer. English consumers now enjoy a variety of innovations and energy services, such as hedged contracts and multi-utility savings offers, which were unavailable before. Thanks to a concurrent (though partly unrelated) shift from coal to natural gas by Britain's power generators, all this has been accompanied by lower emissions of local pollutants and greenhouse gases.

That is the sort of tale that inspired California's early reformers. Unfortunately, the political compromises struck during the Peace Death March led the reformers to embrace a "deregulation" plan that did not take into account some important differences between California and Britain. Perhaps California's reformers were lulled into complacency by the apparent ease with which other markets had liberalized. Europe's deregulation, initially in Britain and Scandinavia and more recently across the rest of the European Union, has not resulted in reliability problems. But credit for that belongs not exclusively to European models of reform but also to excess generation capacity. Europe's top-heavy, state-dominated power sector has tended to "gold-plate" its assets (through higher tariffs paid by captive customers). In contrast, California deregulated into a market that was already showing signs of tight supply, for reasons described in detail below.

California officials also took steps that inhibited the development of a competitive retail market. Rather than allowing prices to fluctuate, politicians decided to freeze retail electricity rates for a few years. That pleased the big utilities, who feared that deregulation might send power prices plunging. As a result, retail consumers were given no incentive to cut power use even when wholesale prices skyrocketed. Ironically, this rate freeze came back to haunt the utilities when wholesale prices shot up unexpectedly in 2000 and 2001: unable to pass along those higher costs, two utilities were driven to financial ruin. The absurdity of this particular aspect of reform in California cannot be overstated. It ensured that people have no sense whatsoever of the true cost of power as they're using it.

Another problem was that the state's politicians agreed to compensate the utilities generously for "stranded assets"—such as the big power plants they had built before deregulation suddenly changed the rules of the game. That sounds fair enough, but California agreed to value those assets much more generously than other states. The Foundation for Taxpayer & Consumer Rights, a leading California activist group, penned this critique of the 1996 deregulation plan:

> Instead of giving consumers the ostensible benefits of deregulation—competition leading to lower prices—the law freezes all utility rates for four years at June 1996 levels, which were approximately 50% above the national average. The freeze requires ratepayers to pay off the costs of "stranded assets" that would otherwise force the utility companies to keep their rates uncompetitively high under deregulation. In effect, consumers are being forced to invest in and underwrite the utility companies so they can compete for big customers.

Worse still, officials decided to burden new entrants to the business with part of the cost of these stranded assets built by the incumbents. In this way and others, newcomers were severely handicapped in their ability to compete on price. Many competitors and potential rivals fled the market—or were strangled by its oppressive rules.

Foes of market reforms insist that such flaws are inherent in deregulation. Yet a number of other states largely avoided making these mistakes. In Texas, utilities were free to enter into long-term contracts in order to hedge against the risk of volatile prices; in California, the rules made it very unattractive for them to do so. That proved a crucial flaw that directly propelled California's two big utilities to financial ruin. And Pennsylvania, like Britain, has had success in spurring retail competition from newcomers. California's version of deregulation allowed none of this, and the upshot is that hardly any Californians switched retail suppliers. The real problem is that what California dubbed "deregulation" did very little to unshackle the power sector from the state.

California's Perfect Storm

Even with its half-baked, half-British model, the state might have muddled along for quite some time. Indeed, as everyone forgets now, California's restructured power market worked quite well for the first two years; prices were stable, and reliability was high. However, several unique forces conspired to create what some have called California's Perfect Storm: growth in demand; fierce opposition to new power supply; and, above all, the politics of pork and populism.

First, growth in demand. As computing power has spread to everything from the manufacture of silicon wafers to toasting bagels, California has defied the pundits who predicted that the digital revolution would result in paperless offices and lower energy consumption. Over the 1990s, demand for power in the state went up by almost a quarter; in Silicon Valley, it rose by some 8 percent a year during the late 1990s. It is not only the power-guzzling "server farms" of the Internet revolution itself that are responsible for the increase in demand, but the wealth it has spread in California, which allows everybody to consume more power. Moreover, the state's energy planners made insufficient allowance for the demographic shifts (including heavy immigration) of recent years. Still, state officials deserve some credit for their strenuous efforts to encourage energy efficiency, without which the state's energy consumption would surely be even greater still.

Despite this rise in demand, Californians have determinedly resisted development of big power plants in recent years. After the last of the nuclear power plants were completed, two decades ago, the construction of big power plants ground to a halt in the state. When San Francisco's local power utility proposed installing a floating power station on a barge to avert blackouts in the summer of 2000 during the early days of this crisis, noisy objections from greens killed the idea. A plan by an independent power producer for a new plant in San Jose, in the heart of the electricity-starved Silicon Valley, was nearly killed off by Cisco, one of the world's largest technology companies and a huge local power-guzzler. Even a proposal to tap green geothermal energy was blocked. Wags now say that California has gone B.A.N.A.N.A.s (build absolutely nothing anywhere near anybody).

Local activism aside, California officials have found plenty of ways to discourage firms from building big nuclear, coal, or gas power plants. The state has long had the toughest environmental laws in America; these made power generation unattractive. Making this worse was the murky and politicized way in which deregulation happened in the state. Thanks to such uncertainty, the utilities built no new conventional power plants at all in the state in the 1990s. That contributed to the problem, but it's still wrong to say that there was an energy shortage in California. While giant power plants were out of fashion, small ones really took off during that decade. In fact, if you added up the output of all of the new micropower—in the shape of "captive" power plants (those built by companies for their own use), solar panels installed by green consumers, and so on—added during the 1990s, it would produce more electricity than all of California's nuclear plants combined.

The real problem, then, was this: peculiar politics. When officials set out to reform the power sector, their stated aim was to introduce competition and deliver lower prices for retail customers. Yet, as the Peace Death March suggests, the way the reforms were designed made this unlikely to happen. Worse still, regulators at the state and federal levels often squabbled and interfered in the electricity market.

Regulators were sometimes needlessly suspicious of market instruments. They actively discouraged utilities from hedging their price risks through long-term contracts or financial derivatives. Since the state's utilities were thus led to buy all of their power on the volatile spot market, they were completely unprotected when wholesale prices skyrocketed in the run-up to this crisis. When market prices for wholesale power increased in response to the summer supply squeeze in mid-2000, panicky officials ordered those prices to be capped. Predictably, this caused an even worse supply squeeze the following winter, because the caps discouraged generators from adding generating capacity.

On the other hand, regulators sometimes naïvely expected the market to sort out other problems of transition all by itself. For example, the Federal Energy Regulatory Commission (FERC), America's top federal electricity regulator, was extremely lax in supervising the wholesale market for power, even as evidence mounted that some power generators may have been manipulating California's flawed system to their own advantage. Indeed, in 2002 federal investigators discovered that a handful of energy-trading companies, including Enron, had tried to "game" the electricity system in California in an attempt to push wholesale prices higher. According to internal company memos that surfaced after Enron's collapse, some traders used covert schemes (with names like Get Shorty, Death Star, and Ricochet) to disguise bogus trades or to overload the grid at choke points and so trigger extra payments for relieving the "congestion" that they themselves had caused.

When confronted by allegations of such shady dealings, Jeffrey Skilling, the former chief executive of Enron, told a congressional panel in 2002 that California's deregulation plan was designed with so many inherent loopholes and contradictions that "the rules weren't quite clear." Self-serving words, to be sure, but there is some truth there too; a vigilant regulator would have made sure that the rules were made clear before the crisis struck. When the FERC finally woke up and looked seriously into this matter in August 2002, it too concluded that while some of those strategies were crooked, others were probably legal. Less ambiguously, some of the tactics deployed by power companies in California were clearly illegal and should have been spotted by regulators. Evidence surfaced in 2002, for example, that, in order to create an artificial scarcity and drive up prices, some generators had shut down power plants during peak periods of the crisis under the pretext that they had broken down.

Such behavior is outrageous but not especially surprising: there will always be those who seek out the loopholes and exploit the weaknesses of every new system. That is why any new set of market rules in any industry must be designed with proper safeguards to deter, catch, and punish criminal manipulation. As evidence from around the world has made clear, deregulation does not mean *no* regulation. In fact, you often need regulators that are more proactive—not less—to ensure that a liberalized market delivers on its promised benefits of efficiency, innovation, and lower prices. That's especially true during the transition to proper competition, when the new rules of the road are not yet set in stone and companies try to get away with as much as they can. The real surprise here is that the FERC under Bill Clinton and the very early days of George W. Bush was not more vigilant in its role as policeman.

This schizophrenic attitude among regulators and politicians toward the market explains why it is wrong to blame market forces for this mess. Michal Moore, who served as a commissioner on the California Energy Commission during the crisis (and who was one of the few sane voices in this debate), summed it up beautifully: "What we have in California is not a market failure; what we have is regulatory failure."

Always the Special Case

Postcrisis conventional wisdom in America holds that energy and markets simply do not mix. One popular argument is that energy is simply too critical an input into the economy to be trusted to the vagaries of the market. That sounds like a plausible theory. After all, any modern economy would grind to a standstill without ready access to energy, as Britain and parts of continental Europe discovered during the gasoline riots of 2000. Arguing against that is the fact that interstate trucking, banking, and even food distribution—all industries subject to vigorous competition—are critical to the economy.

Water is an even better counterexample, as it is closer to the sort of commodity (along with power and telephone service) that used to be thought of as too essential to allow private-sector participation. Yet look around the world and you find that dozens of countries (though not the United States), from Mexico to Argentina to the Philippines and China, have recently introduced reforms in the water sector that encourage private investment and management. Even in statist France, water has been provided by private-sector companies for more than a century. One such firm, Suez, provides water and electricity services in more than 130 countries in the world.

Those concerned about the plight of the poor and elderly in society argue that free markets are too volatile and painful for essential commodities. Actually, competition is more likely to produce lower prices that would benefit such people the most. Besides, the better way to protect the poor from price spikes is through targeted subsidies designed to help only the needy—not wasteful universal subsidies (which do not distinguish between a billionaire like Larry Ellison, the boss of Oracle, and Tassie Dykstra) and blanket state control. Electricity-market reforms in places like Europe typically include a safety net of targeted subsidies that help the "fuel poor." For example, London Electricity installs sensors in the homes of especially vulnerable customers—like impoverished old people living alone—that alert the company when temperatures in those homes fall below a dangerous threshold.

Yet another argument offered is that electricity is too technically complicated and asset-intensive a business to be trusted to free markets, which tend to focus on short-term results. Again, there is something to this: it is true that electricity (unlike grain or automobiles) cannot be stored easily, and so some sort of regulated grid operator is necessary to match demand and supply perfectly at all times. It is also true that big power plants (though not micropower plants) take several years to bring on-line, so the natural supply response to shortages or price spikes can be delayed. In other words, energy markets are not perfect "textbook" markets.

But does this more nuanced argument really add up to a convincing case for monopoly or government control? After all, other complicated and asset-heavy businesses, such as telecommunications and airlines, have been successfully deregulated around the world. Even in electricity, some trailblazing countries have met with success—as the various examples noted by the IEA above attest. By maintaining a vigilant but carefully circumscribed role for government regulators in such areas as antitrust scrutiny, these countries have been able to do what seemed unthinkable just a few years ago (and, to folks like Loretta Lynch, still seems unthinkable): introduce market incentives into the generation, transmission, and retail distribution of energy services.

In short, market reforms work when done properly—but, as with most worthwhile endeavors, they carry pitfalls. Paul Joskow, an MIT professor and a leading architect of electricity deregulation, summed up the challenges: "I sometimes think people involved during the last four years haven't gotten the message that this isn't a piece of cake . . . It's not something that you can snap your fingers and say you have a competitive market."

From Carter's Cat to Bush's Vampire

If market reforms are hard to get right, and disastrous when they go wrong, why bother at all? The answer to that reasonable question points to the real lesson to be learned from the California crisis: electricity is simply too precious a resource not to use wisely—and properly supervised, competitive markets can play an important part in that wise use.

Most people in the world think of electricity as an entitlement without limit or cost. But power is produced and transmitted at great expense, even before the damage done to the environment or human health is accounted for. California's power crisis has already provoked a national debate on energy policy. It may even have provided a wake-up call to Americans, the world's biggest guzzlers of fossil fuels, to think seriously about how they use energy.

The last time America had a real debate about energy was after the oil shock produced by the Iranian revolution in 1979. Unfortunately, the memory that lingers from that era is of Jimmy Carter in a shabby cardigan sweater lecturing Americans to turn down the thermostat. Carter tried to toughen up his image, insisting that America needed to declare "the Moral Equivalent Of War" on oil producers. It did not take critics long to pounce on his conservation-based approach, cruelly dismissing his roar as a mere M.E.O.W. As Carter faded from the scene, a decade-long period of low and stable energy prices followed—and Americans quickly forgot about conservation.

Ever-rising demand for electricity cannot be met forever by building more and more power plants, as Dick Cheney seemed to think when he mocked conservation as a basis for a national energy strategy. The grand energy plan unveiled by the Bush Administration in May 2001, for example, put forward the dubious proposition that the only way out of America's energy "crisis" was to build hundreds upon hundreds of new power plants. As most energy experts acknowledge, a sensible energy strategy must aggressively tackle demand too. That is not to say, however, that conservation and warm sweaters will save us all, as some greens would have you believe.

The mistake that most people make on both the left and right is in confusing two very different sorts of demand-side approaches: conservation and energy efficiency. Conservation is the intentional use of less energy (say, by turning off some lamps or radiators), which always means less of the good things that energy brings (like light or heat). Energy efficiency, in contrast, means squeezing more of the good things that energy makes possible out of every ounce of fuel and every appliance. The former, like Jimmy Carter's cardigan, usually involves some discomfort and sacrifice but sends the moral signals that liberals tend to like. Energy efficiency, in contrast, does not feel so warm and fuzzy. However, while conservation may or may not be a good idea (should Granny really turn down the thermostat on a cold winter night?), efficiency is always sensible. This is especially true in profligate America, which consumes more energy per person than just about every other developed country in the world.

So how do you encourage energy efficiency? One popular but controversial approach is to stiffen energy-efficiency standards for heavy appliances, cars, and the like. The best example is the development of power-sipping refrigerators: a new fridge today uses half the power of a comparable model from 1986, and it costs the owner half as much to operate. Industry usually hates such regulations and often finds a sympathetic ear among conservatives: in 2002, for example, opposition led by Republicans thwarted efforts to stiffen the Corporate Average Fuel Economy (CAFE) law, the controversial statute that dictates fuel-efficiency standards in cars and trucks, while the Bush White House weakened standards proposed by Bill Clinton on air conditioners. Businessmen always insist that they cannot meet higher standards without big price hikes or massive investments in technology; almost always, such protests are overdone. However, sometimes the industry in question is right about the unfair burdens. And if such regulations resulted in higher prices, the poorest consumers would be hit the hardest. That points to why such efficiency standards must not be imposed willy-nilly: they are blunt, second-best measures that rely too much on the judgment of bureaucrats.

Market forces offer a better (if not always politically popular) way to spur efficiency: introduce proper price signals. Virtue alone will not spur very much efficiency or conservation, but financial incentives might. One of the deepest flaws in California's "deregulation" was the fact that retail tariffs were fixed even as wholesale prices were freed. Having experienced no retail price fluctuations, consumers had no incentive to reduce consumption, no matter how high the wholesale price. Regulators eventually did raise rates, but the new rates were fixed too: they did not fluctuate in real time with the wholesale price, and so did not send price signals. Only introducing "smart meters" and "real-time pricing" for all customers would do that.

This may sound like an arcane technical matter, but in fact it could make a huge difference to the electricity bills of people like Tassie Dykstra. The Electric Power Research Institute (EPRI) conducted a fascinating study that suggests that a drop in demand of only 2.5 percent during a power crunch in California could reduce the wholesale peak price by up to 24 percent. The reason is that the price for power skyrockets during peak moments as power-distribution companies desperately bid up the price of scarce electrons from power suppliers in order to keep their customers from suffering blackouts. Nudging down demand by just a few percentage points, concluded EPRI, effectively splashes cold water on the overheated market, and so drives down prices by a disproportionate amount.

But how exactly can consumers be persuaded to cut back a bit on their power guzzling during peak hours? Well, they won't if prices are firmly fixed regardless of demand, as they are in most non-liberalized markets. Airlines don't charge a fixed fare for all seats on all planes going to a particular destination, regardless of whether one flight (the one in the middle of the night) is half empty while another (at the end of the business day) is overflowing beyond capacity. Of course, airfares vary based on the load on the system—and so too could electricity rates if only the industry and regulators were not so locked into a 1950s mentality.

Severin Borenstein, a professor at the University of California who is the leading expert on the state's electricity system, argues passionately for the introduction of "dynamic" pricing that would provide users of electricity with essential information that they are denied today: the true cost of the power that they use. With smart meters, which are practical using even today's off-the-shelf technology, they could easily provide the few percentage points of relief that would keep the system from overloading. These meters (or even a crude version found in some European countries—"red light/green light" meters) could produce such savings because consumers would be informed when the system nears overload and rates spike. As millions of individuals or their Internet-savvy appliances respond to the signal (say, by washing clothes at night instead of during the sweltering daytime peak), the spike is diminished—and prices reduced considerably.

California's crisis has spurred moves in Europe and America to improve the efficiency of public-lighting fixtures, with the help of bulb manufacturers like Philips. Home Depot, a giant home-improvement retailer, reported in 2001 that energy-related products (such as efficient lightbulbs, glazing for windows, insulation, and so on) were flying off the shelves in parts of America with high energy prices, not just in California. The newfound awareness of efficiency also led to efforts to reduce the vast quantities of power wasted by appliances as they wait in standby, or "vampire," mode. It turns out that seemingly dormant televisions, fax machines, cellphone chargers, and the like waste huge quantities of power while standing by. This adds up: in Germany alone, the wasted energy equals the output of a nuclear power plant, according to the International Energy Agency. The IEA has formed a coalition of manufacturers to agree on standards and reduce standby power levels without affecting the performance of appliances.

Even George Bush got into the act in his own inimitable way: he baffled audiences during his monthlong vacation in August 2001 by proclaiming his support for "vampire-busting devices." He had been informed that the average American household wastes between 4 and 7 percent of its power consumption this way, and so had ordered federal agencies to investigate ways to fix the problem. Environmentalists were tickled pink that Bush and Cheney were forced to swallow their oilman's pride and add a number of such measures aimed at conservation and energy efficiency to their final energy plan.

From Exploding Manholes . . .

Thank goodness, for the Bush Administration's initial premise used to justify industry-friendly policies was a preposterous one: that America was mired in an energy supply crisis. On the contrary, as the record number of power plants planned in 2001 (partly in response to claims of shortage arising from the California crisis) attested, there was every sign of impending overcapacity in power generation. There was no shortage of oil or gas in the world either, as chapter 4 makes clear. Indeed, *Barron's* ran a cover story in mid-2001 trumpeting "The Coming Energy Glut." In short, there was never any real energy crisis.

Nevertheless, people continue advocating policies that, in effect, would provide massive subsidies for energy producers. They argue that protected federal lands must be opened up for drilling too, lest the "crisis" worsen. There may well be a good case for examining this matter on a case-by-case basis, weighing the legitimate trade-off between environmental benefits and energy resources. However, such a sensible approach certainly was not evident in the debate over Alaska's Arctic National Wildlife Refuge (ANWR). In the most absurd twist, some proponents of opening up part of this protected reserve to oil drilling invoked the California power crisis as justification—despite the glaring fact that oil is hardly used in power generation in America (not to mention that the ANWR oil would not get to market for another decade anyway).

If the sorts of policies initially advocated by Dick Cheney's special task force on energy policy in 2001 were really put in place for any extended period of time, the results would be most unfortunate. By encouraging big nukes and big coal, America would be locking in old, dirty, inflexible, and inefficient technology for decades to come. By doing so, the country would choke off at birth such innovations as micropower, and so forgo the chance to leapfrog ahead to the age of the Energy Internet.

However, despite its deficiencies, the Bush energy plan deserves credit for highlighting one awkward problem that environmental groups typically try to gloss over: bottlenecks that keep oil, gas, and power from getting to the end user. In the words of a panel of energy experts commissioned by the Council on Foreign Relations and Texas's Rice University, there is a serious "energy infrastructure" crisis looming.

Officials have reformed the wholesale and retail power markets but neglected the grid. Ever-increasing environmental regulations, red tape, and the B.A.N.A.N.A. syndrome have also made it nearly impossible to build new transmission lines, refineries, or any other essential bit of energy hardware. Worse yet, nobody has a clear incentive to maintain and upgrade the energy infrastructure used by all.

We have California to thank for shining a light on the infrastructure problem. As wholesale trading expanded, bottlenecks built up in various places. The best example is Path 15, the chief grid connection linking the northern half of California with the southern half. Blackouts have sometimes occurred simply because power was choked off. A similar choke point kept cheap power from Maine from getting to Massachusetts on a particularly hot August day in 2001, when consumption hit a record level; enough power for 150,000 Boston homes had to be purchased at sky-high rates elsewhere instead. The grid deserves blame for another eye-catching development that sweaty summer: exploding manholes throughout Washington, D.C. As decaying and overloaded power lines blew up underground, they sent manhole covers in Georgetown and Dupont Circle flying tens of feet in the air. Surely the richest country in the world can afford better infrastructure than this.

. . . to Windmills in the Sky

So what's the best way to bust these bottlenecks? Curiously, Gray Davis and George W. Bush, bitter enemies though they were, saw eye to eye on this question: state intervention. Davis came up with a blunt fix for Path 15 and other California choke points: he wanted the state to control the grid. Bush proposed that the federal government take on sweeping powers of "eminent domain" to push through new power-transmission lines and other essential bits of energy infrastructure. Such proposals angered conservative states' rights advocates as much as green lobbies, but there is sense in them: there is a justifiable government role in ensuring adequate investment in the transmission system.

Even so, a market-based approach would be a much better way of fulfilling that government role than wielding the bludgeon of state control. Rather than encourage state takeovers of America's balkanized grid, regulators would do better to give private grid operators strong incentives to upgrade the system. In this, too, California seemed to have had a positive (if unintended) effect. In July 2001, much to the shock of utilities and state regulators, the FERC boldly ordered the country's many grid operators to start consolidating along regional lines. Perhaps predictably, many critics—especially public-power advocates—howled that the private sector simply cannot be trusted with something so vital to the national interest as the electricity grid.

The best reason to think otherwise comes from Roger Urwin, the boss of Britain's National Grid, which runs the power grid in England and Wales. His firm used to be a state monopoly, and he was a top manager in the electricity business in that era. Back then, he recalls, he too was skeptical of market reforms. However, after reforms turned National Grid into a regulated private-sector monopoly, it managed to post healthy profits for years, provide new and innovative services, and even expand into the American market. Here's the clincher: it did all this while investing perhaps ten times more money to upgrade England's grid (once adjusting for its much smaller size and infrastructure base) during the 1990s than all grid operators across the Atlantic combined invested in America's decaying power grid during that decade. What explains National Grid's remarkable turnaround—given that we are discussing the same company, with the same grid assets, and even the same veteran manager? Market reform has transformed his firm from a sleepy, quasi-governmental agency into an aggressive firm accountable to shareholders. "The crucial difference," he said with pride, "is culture."

America should certainly create incentives for grid investment, but an even better way to get around grid bottlenecks is to expand micropower. By placing small units close to the end users, utilities (or end users) need not send so much power down those aging lines in the first place. In the past, utilities sometimes used high "standby" charges or bogus safety worries to obstruct micropower. No longer, if California's energy czar gets his way. The state decreed in early 2001 that all "standby" charges are to be waived for the next two years for micropower units. Micropower technologies, ranging from fuel cells to microturbines to solar panels, have taken off there.

Moral support from the energy czar certainly helped, but the biggest reason micropower took off in California was that power prices in the state were so high. And that pointed to the inherent risk in putting the industry under state control: the czar's newly created Power Authority locked in very high electricity prices for a decade or more through long-term contracts negotiated at the peak of the power crisis.

Yet the lessons of that fiasco seemed lost on advocates of public power. Though Medea Benjamin and her group lost the referendum to put San Francisco's power supply under government control, she remained firmly committed to her cause. The Power to the People coalition, in which Global Exchange was a partner, continued to argue:

> It has become clear that we cannot leave such a critical resource as energy in the hands of companies that put profit before the social good. We need public ownership of our energy. Public power is not a new or radical concept. It already exists in the state of Nebraska, over 2000 cities in the United States and 31 municipalities in California. Here in California, in the municipalities where public power programs already exist, it has provided, on average, rates that are 20% lower than investor-owned utilities, and better programs for conservation and the use of renewable sources of energy. Our vision is one of locally controlled

> municipal utility districts, coordinated by a state-wide Public Power Authority. By taking out the profit motive and adding local control, public power would put the needs of Californians and our environment front and center—where they should be!

It is true that there are isolated examples of well-run and even innovative "munis," foremost among them the one that supplies power to Sacramento. This particular utility even helps customers acquire solar energy systems, which cost a lot to buy but little to operate, by offering to bundle together financing for them into their home-mortgage payments. But such innovation is the exception, not the rule. Most other public-power firms around the world have done little to promote innovative micropower technologies like fuel cells. And the claim that rates have been 20 percent lower at California munis than at the state's big utilities is pretty underwhelming when you consider how bloated and inefficient those monopolies have been. It also sidesteps the proper test: How would munis fare against utilities run by professional managers in a competitive market? The record of utility monopolies, whether run by conventional managers or by elected public-power advocates, is pretty poor. Even residents of Sacramento are smarting from their celebrated muni's costly and ill-fated foray into nuclear power.

What is more, the experience outside the Disneyland economics and politics of California clearly suggests that market reform offers the best hope of putting micropower in backyards. Or, as residents of the Windy City will soon find out, on their rooftops. Some of the biggest converts to micropower have been the heads of cities and states across America. Mayor Richard Daley, for example, has vowed to turn Chicago into "the greenest city in America" by getting a fifth of its power from green sources by 2006. Experts say that as much as half of that green power could come from supermodern windmills, whirring silently inside cages encrusted with solar panels, which are to be installed atop the city's buildings.

How is this possible? Part of the explanation lies in the nifty advances in the technology behind renewables, which are getting cheaper year by year. However, the bigger part of the explanation lies in the fact that market reforms are now shaking the lazy utility business out of its decades-long slumber. Daley and nearly fifty other mayors got together in 2000 and decided to use their collective might to tap the country's evolving electricity markets. Bill Abolt, at the time Chicago's environment commissioner, explained: "We decided we want power that is cleaner, cheaper, and produced close to home . . . and deregulation made it possible."

Hands On, Hands Off

Medea Benjamin is right in a way: electricity is simply too "critical" and "life-giving" to squander. California's crisis, and its bungled handling of that crisis, have reminded people everywhere to value power properly and to use it efficiently. The world must now decide whether the best way to do that is to create an innovative and genuinely competitive market with a vigilant regulator and proper price signals—or through the tried and tested stagnation of state control.

It would truly be a pity if politicians around the world mistakenly blame deregulation for California's muddle. In principle, the state should merely serve as a textbook case of how not to do things. In practice, though, reformers may find that California's debacle makes it much harder for them to overcome the many vested interests opposed to competitive markets—ranging from coal unions fighting for subsidies to incumbent utilities defending cozy monopolies to cosseted energy giants and energy cartels (like the ones described in the next chapter, on oil) fighting innovations such as fuel cells. The real solution to America's halfhearted attempt at electricity reform is clear: more deregulation, not less. It took Britain the better part of a decade, with some anxious moments along the way, to get its market reforms right.

Any worthwhile endeavor carries risks, and the experience from around the world suggests that it is neither foolhardy nor impossibly difficult to make reform work. Regulators and politicians must speed ahead—but alert and with both hands firmly on the steering wheel, not asleep behind it. Only then will the world's electricity networks be transformed into the vibrant Energy Internet worthy of the twenty-first century.

4

Oil—The Most Dangerous Addiction

A FANTASTICAL PLEASURE DOME rises up out of the sands outside Riyadh, the capital of Saudi Arabia. The giant castle is surrounded by manicured gardens and flowering plants of the sort that God never intended for the Arabian Desert. Hundreds of exquisitely groomed camels frolic about the vast grounds. The owner, a favorite prince of the ruling family, was throwing a sumptuous banquet for foreign dignitaries. The guest of honor—Ali Naimi, Saudi Arabia's minister of petroleum—was the custodian of the gooey stuff that makes such opulence possible. Even Bill Richardson, then America's Secretary of Energy, took the trouble to attend.

How long will the black gold that financed all this continue to ensure this prosperity? Could the turbulence of global oil markets at the end of the twentieth century mark the beginning of a new energy era? Are we in the dying days of the Age of Oil? Naimi did not hesitate before replying. "I am not in the business of forecasting or dreaming," he said with a wry smile, "but I am certain of one thing: hydrocarbons will remain the fuel of choice for the twenty-first century."

That view put Naimi at odds with a growing chorus that says that oil's grip on energy markets may soon start to weaken. Among them is the best known of Naimi's predecessors, Sheikh Zaki Yamani, whom the Western world came to know as the face of the OPEC oil cartel during the turbulent 1970s. The prospective scarcity of oil, such people argued, combined with the instability of undemocratic Arab regimes, will cause a steep rise in hydrocarbon prices over the next two decades. The voracious demand for oil in the developing world will make things even worse. The argument goes that all this will lead to chaos unless Western governments take drastic action to bolster "energy independence."

Even George W. Bush's noisy campaign to oust Saddam Hussein from power grew inextricably linked with such concerns. Though Bush insisted he was pushing for Saddam's ouster only because of worries about weapons of mass destruction, it did not escape anyone's attention that Iraq happens to have a bit of oil: the biggest reserves of conventional petroleum in the world outside of Saudi Arabia, to be precise. Yet thanks to more than a decade of UN sanctions and Hussein's own misrule, Iraq was producing less than 1 million barrels of oil per day in September 2002—less than a third of its previous peak and a tenth of the output of Saudi Arabia. If a post-Saddam government lifted Iraq's production to a level consistent with its reserves, argued some strategies, the country could become a major force on the world oil markets.

For Americans concerned about overreliance on OPEC oil, this sounded pretty attractive. If a pro-Western regime turned its back on OPEC and threw open the taps, went the argument, it might even undermine Saudi Arabia and the oil cartel altogether. However, there was a flaw in that fantasy: liberating Iraq's oil is sure to prove much harder than liberating the Iraqi people. Saddam Hussein's regime had damaged the country's oil infrastructure even before postwar looting wrecked it altogether. Industry experts and veteran oilmen agreed that it could take up to ten years for Iraq to become a serious rival to Saudi Arabia. A democratic Iraqi regime would probably still decide—in its self-interest—to remain inside OPEC and collude with Saudi Arabia to fix prices rather than turn itself into America's filling station. After all, that's what (semi)democratic Venezuela and Nigeria do from within OPEC. Mexico and Norway, non-OPEC democracies that happen to produce a lot of oil, also collude with the Saudis from outside the cartel. In other words, "regime change" in itself need not transform the global oil market. Cambridge Energy Research Associates, an industry consultancy, put it this way: Iraqi exports will go up now that Saddam has been ousted and UN sanctions lifted, but that does not mean "a massive, rapid increase in production that will depress prices, displace other Gulf producers, and render OPEC impotent."

In short, the second Bush to take on Saddam Hussein will probably be long gone from the White House before the oil markets are transformed by Iraqi oil. You shouldn't write off Saudi Arabia or OPEC just yet. However, here's one thing you can be sure of, especially if the troubles in the Middle East continue for some time: oil prices will continue to swing wildly for years to come. Hang on to your hats!

A Roller Coaster Without Brakes?

The wild gyrations of the oil market at the end of the twentieth century caught many observers by surprise. After more than a decade of relatively stable oil prices of around $20 per barrel during the 1980s and early 1990s, they collapsed to around $10 in 1998, only to soar to a ten-year high of more than $35 a barrel in 2000. That spike set off a political crisis over gasoline prices and shortages in America's midwestern states, and it ensured that energy became a hot issue in the presidential campaign between George Bush, the Texan oilman, and Al Gore, author of the environmentalist tome *Earth in the Balance*.

Soaring prices also provoked riots by truck drivers, farmers, and other heavy gasoline users, paralyzing several European countries. Worried that Americans might react in similar fashion to rising heating-oil prices that winter (and no doubt mindful of the impending presidential election), Bill Clinton took the extraordinary measure of releasing some oil from the government's Strategic Petroleum Reserve. Yet oil prices remained buoyant during the early days of the Bush presidency. In a sense, that suited America's new leader just fine. After all, oil companies—especially the smaller, less efficient firms led by Bush's friends back in Texas—benefited from high prices. What is more, the price spike suited Bush's ideological agenda. Capitalizing on the fears surrounding those high prices, he put together a supply-side energy policy that assumed America was mired in a serious energy crisis.

Most such voices had faded from the scene after the forecasts they made during the oil shocks of the 1970s—impending oil shortages, ever-rising energy prices, even economic collapse—proved to be baseless. Energy doom-mongering is once again in fashion. Several arguments are particularly popular among the born-again naysayers. One camp argues that recent price spikes are signs that the world may be entering a new energy shock. Bush and his supporters are clearly sympathetic to this view. The second camp reckons that the real danger exposed by the recent price roller coaster is the revival of the once-moribund oil cartel. They are convinced that Naimi and his OPEC brethren are set on strangling the rest of the world with ever higher prices while they frolic on their vast estates. Others go further: they see the recent bumps as a warning that the world is beginning to run out of oil. According to this argument, the coming shock will not be short-lived, as it was in the 1970s, and oil prices will surge ever higher as reserves are quickly depleted. Take any one of these arguments to its natural conclusion, and the world is already in the early stages of a serious energy crisis. And the more accurate the doomsayers are, the sooner the hydrocarbon era will draw to a close. That would make a mockery of Naimi's assertion that oil will remain the "fuel of choice" for this coming century.

Is the kingpin of OPEC really so wrong? The short answer is no—and yes. Naimi is right, and the pessimists are wrong, in one important sense: there is no real energy crisis. In fact, there is good reason to think that oil is plentiful today and likely to remain so for several decades to come. However, Naimi's insistence that oil will retain its grip on the world economy indefinitely may still be wrong. Scarcity is not the only reason that the world could shift away from oil. Increasing price volatility, oil's environmental impact, and worries about the growing power of Middle Eastern producers have already spurred a global search for alternatives to the century-old combination of petroleum and the internal combustion engine. In time, advances in such technologies as fuel cells and hydrogen-based energy could provide a clean alternative that breaks oil's near-monopoly grip on transportation. As Sheikh Yamani is fond of saying these days, no doubt in part to torment his successor, "The Stone Age did not end for lack of stone, and the Oil Age will end long before the world runs out of oil."

What Energy Crisis?

The most extraordinary aspect of George Bush's energy policy may be that it pays homage to Jimmy Carter. As soon as Bush took office, he declared that he was "deeply concerned" about an energy crisis: "It's becoming very clear to the country that demand is outstripping supply." He asked Dick Cheney, his hard-charging Vice President, to come up with an energy plan that would work out "how best to cope with high energy prices and how best to cope with reliance on foreign oil." Both Bush and Cheney constantly reinforced the theme that their energy plan would ensure that Americans' God-given right to cheap energy was not compromised.

The echoes are unmistakable. Here is Carter's energy plan from 1977: "The diagnosis of the U.S. energy crisis is quite simple: demand for energy is increasing, while supplies of oil and natural gas are diminishing. Unless the U.S. makes a timely adjustment before world oil becomes very scarce and expensive in the 1980s, the nation's economic security and the American way of life will be gravely endangered." Despite all the political grandstanding and the screaming headlines, there is no good reason to think that the world is entering anything resembling an energy crisis. Comparisons with earlier oil shocks are vastly overblown. For one thing, the causes of recent energy crises are quite different from those that produced the oil shocks of the 1970s and the lesser turbulence during the Gulf War.

Opportunistic following the Iranian revolution, OPEC now tries hard to stabilize energy prices. The cartel's ill-disciplined ways make it difficult, but its stated aim is to keep prices between $22 and $28 a barrel. The leaders of big consuming economies like the United States have indicated that they can live with that.

What is more, oil consumers have become less vulnerable to oil shocks. The higher fuel taxes embraced by Western governments (least enthusiastically by the United States, it must be noted) in the wake of the earlier crises have done much to spur conservation, fuel switching, and improvements in energy efficiency. Oil's share of industrial countries' imports, and their economies' reliance on it, has shrunk significantly as a result. In the last three decades the amount of oil consumed per dollar of economic output (in inflation-adjusted terms) has fallen by almost half in the rich world. The shift from heavy manufacturing to service industries and especially to information technology has also accelerated this trend.

The earlier energy crises came when the Western economies were soaring and inflation was already high. Higher oil prices quickly led to demands for higher wages—thus stoking even higher inflation. In contrast, the most recent oil price increases came at a time of low inflation in Europe and America, never mind Japan (which has suffered from deflation), and the damage done to the real economy was much less.

The specter that has so spooked some energy experts—the return of $30-a-barrel oil—is not really that scary, looked at historically. Once adjusted for inflation, that "high" price of oil is still below the level reached in 1981. Even the shockingly high pump price of $2 a gallon of gasoline, which sends motorists and politicians into fits in the United States, is not much different, in real terms, from what those motorists' parents paid to fill up their cars three decades ago.

In short, this is no crisis. However, it may *feel* like a crisis to some because the price surge came after nearly a decade and a half of stable, low oil prices. That golden period was the by-product of the OPEC cartel falling into disarray for many years. Its members were simply unable to engineer painfully high prices. The cartel's low point probably came at a summit in 1998, when it decided to raise production just as the economic downturn in Southeast Asia was unfolding. That ill-advised decision flooded the market with unwanted oil and sent prices tumbling down to $10 a barrel.

Even before that collapse, the low and stable oil prices of the 1990s proved a gift for the consuming economies of the world. Indeed, Andrew Oswald, an economist at Britain's University of Warwick, argues that this exceptional circumstance deserves far more credit for America's "miracle economy" of the '90s than the conventional explanations of productivity growth or information technology. Philip Verleger, an oil economist affiliated with the Institute for International Economics, cites evidence suggesting that the pain inflicted on consuming economies by oil price spikes greatly outweighs the benefits seen during periods of low prices. One thing is clear: the exceptional stability of real oil prices during the 1990s undoubtedly lulled the world into complacency. Oil remains a political commodity.

The Opaque Cartel

If you want to sneak a peek into the murky workings of OPEC, head down to the Bunker. Set amid the architectural treasures of Vienna, the squat modern building that houses the OPEC headquarters is, well, ugly. However, its designers did not have aesthetics in mind when they built it. They wanted, and got, an impenetrable fortress designed to protect some of the most powerful and controversial men in the world.

When changing market conditions force OPEC ministers to revise their production quotas, they will often hold a meeting here. The real action begins several days before the official meetings, however, as the ministers arrive. Given the hyper-politicized nature of this cartel, there are endless face-to-face negotiations between individual countries and among various factions. Price hawks like Libya and Iran usually argue for keeping output tight, Western economies be damned; Saudi Arabia and Kuwait, with close ties to the United States, usually push for a more moderate line on prices. But nothing is ever predictable with this bunch. During one tense impasse, Bill Richardson, America's Secretary of Energy under Bill Clinton, cornered a group of ministers by conference call in a hotel suite to make his case for more oil. Such haggling often continues late into the night as alliances are formed and betrayed and countless bottles of Chivas Regal and packs of Dunhill's polished off.

During this whole process, hundreds of journalists from every corner of the world lie in wait, eager for any morsel of information about production quotas or price targets. The slightest tidbits are immediately relayed by satellite to the world's financial wire services. Many of the hacks hang around the lobby of the InterContinental, the favored hotel of several important delegations. As ministers scurry between secret meetings, they are ambushed by journalists eager for a sound bite. Venezuela's diminutive oil minister of the day, Ali Rodriguez, was once nearly knocked over by overeager television crews thrusting microphones into his face.

The oilmen clearly bask in the limelight. The cleverest of the OPEC ministers will sometimes nudge oil prices by making the vaguest of Delphic proclamations. Journalists can also get in on the act. During one particularly long and fruitless stakeout in the InterContinental's bar some years ago, a few British journalists decided to play a prank. One among them, who spoke Arabic, dressed up in a sheet and tied a tea towel on his head. Several similarly attired companions accompanied him as he grandly marched out of the elevators through the foyer, barking orders at them in Arabic. Within moments, several Japanese journalists had leaped to his side, hanging on his every word as if he were Moses himself coming down from Mount Sinai.

Often it is in the late hours that the horse-trading ends and the OPEC heavyweights reach a consensus. The secretariat makes a public proclamation about the new production quotas, and the world oil markets respond on cue. Whenever gasoline prices rise, as they did in 2000, American consumers blame greedy Big Oil and Europeans grouse about government taxes. Just imagine how much more outraged they would be if they realized that the price hikes are really due to such deals struck in the Bunker.

This highly scientific method of fixing the oil price points to the second big concern raised by pessimists: OPEC itself. Geostrategists and energy-security experts fret that recent price surges mean that the cartel is back for good. They worry that most OPEC members remain openly hostile to the West (especially to the United States), and that they may keep oil prices high enough to strangle Western economies. Worse yet, OPEC may be tempted to use oil as a political weapon, as it did in the 1970s.

This camp of petro-pessimists also warns that in a few decades, as non-OPEC oil reserves are depleted, the world will grow ever more reliant on the Middle East. Saudi Arabia and Iraq sit atop over a third of the world's proven reserves of conventional oil. Long after affordable oil stops flowing from the North Sea, Alaska, and even the fledgling fields of the Caspian, those two countries will have lots of cheap oil left—and they may try to charge extortionate prices for it. That long-term concern is legitimate, and as good a reason as any for Western governments to invest now in research into substitutes for petroleum. Indeed, there is reason to worry about terrorists someday taking over those vast Saudi oil fields.

Geopolitics could lead to such a nightmare scenario, but economics do not suggest that a sustained OPEC squeeze is in the cards, for three good reasons. First of all, cartels are devilishly tricky to maintain for any length of time, thanks to what economists call the "free rider" problem: though collectively all members benefit from strict adherence to production quotas, each individual member has a powerful incentive to cheat. By producing, on the sly, just a little bit more than allowed, a particularly greedy or desperate member can pocket huge extra profits and gobble up market share. As they do so, however, other members get irritated and start doing the same—and a vicious circle develops that ultimately leads to a price collapse. That has in fact happened to OPEC a number of times in the past—just as it has to almost every big cartel in every industry.

The only real exception to this rule is the De Beers diamond cartel, which for decades has managed to peddle intrinsically worthless pieces of carbon for outrageously high prices. The cartel's brilliant advertising has bamboozled generations of bachelors, but the most significant reason the cartel has survived is its manipulation of production and stocks. The key is sheer size: De Beers itself controls more than half of the worldwide production of diamonds and so can move the world market at will. That makes it easy to discipline the small fry in its cartel. The oil situation is more complicated, and it makes sustaining discipline difficult. Saudi Arabia's share of production is smaller than De Beers's, making it much less dominant in its respective cartel (though this will change over the next two decades). Furthermore, the prickly politics of the Middle East mean that the Saudis cannot simply issue orders from on high. They must constantly cajole, coddle, and coerce their brethren to agree to specific quota changes—and to stick with them. That is why concerted action by OPEC never lasts.

There is also reason to dismiss the specter of a malicious OPEC out to get the West. It has nothing to do with goodwill: many of the cartel's members would indeed be delighted to see the Great Satan suffer in their oily clutches. That list could include even Venezuela, an old American ally, thanks to the rise to power of Hugo Chavez, the country's Yanqui-bashing ruler. No, the reason to think that the cartel will not sock it to the West as it did in the 1970s is an even more powerful impulse than hate: greed.

History shows that the biggest losers from the Arab embargo and other energy shocks of the past were not Western consumers, but the oil-producing economies themselves. That is not to say that the West did not suffer; however, it adjusted to the oil shocks and ultimately emerged more energy efficient. The OPEC countries, in contrast, lost billions of dollars in potential revenue as the moribund consuming nations cut back sharply on oil imports. OPEC will find it nearly impossible to sustain a policy of painfully high oil prices over the long term. This is particularly true for Saudi Arabia, whose current rulers are heavily dependent on American military and political support; they are keenly aware that they must not kill the goose that lays their golden egg. If a rabidly anti-Western figure like Osama bin Laden were ever to control the Saudi spigots, however, that goose is probably cooked.

There is another reason to believe that OPEC will not be able to squeeze the world economy by raising prices ever higher: market forces. If the cartel sustains prices at a painfully high level, companies and governments will rush to develop non-OPEC supplies of hydrocarbons. Those supplies would act as a safety valve that places an upper limit on OPEC prices. This is precisely what has happened whenever OPEC has tried to engineer high prices in the past: private companies have flooded the market with oil from marginal fields in the Gulf of Mexico, technically complex fields in the Alaskan tundra, expensive fields in the North Sea, and so on. Though it will strike some readers as counterintuitive, that's also the reason consumers should stand up and cheer when they hear of yet another merger between oil majors such as the recent ones that put together Exxon and Mobil; BP and Amoco and Arco; Total and PetroFina and Elf Aquitaine; Chevron and Texaco; and several others too: because they alone have the deep pockets required to develop the remaining non-OPEC fields, these giants provide a useful check on the cartel's market power.

In Praise of Big Oil

The recent wave of super-mergers may please oil bosses and institutional investors, but it has sparked widespread alarm in most other quarters. Indeed, bashing Big Oil is a popular sport. Soaring oil prices and high gasoline taxes helped inspire widespread street protests across Europe in late 2000. Though the initial target was government, the oil majors inevitably came under attack from all sides. Oilmen met with rough treatment during the American presidential election campaign that year as well. Vice President Al Gore took some shots at the price fixers at the OPEC cartel, but the lion's share of his populist attacks was aimed squarely at big oil companies. His campaign cried foul when the Chevron takeover of Texaco was announced: "Given the fact that oil companies saw their profits rise by over 300 percent in the past year, it raises the question whether big oil is getting too big." Rather than offering a vigorous rebuttal, George W. Bush managed only timid retorts that the real culprit was "big foreign oil"—in other words, not his friends back in Texas.

But is the transformation of Big Oil into Enormous Oil really so worrying? The only legitimate question that should concern regulators—are consumers harmed by these deals?—is straightforward to assess and, if necessary, to remedy. The downstream end of the oil business that most affects consumers, be it refining or retail marketing, is relatively transparent. What is more, the painfully low margins typical of the retail end of the oil business are evidence of intense and increasing competition (including from independent refiners like Tosco and supermarkets like Wal-Mart peddling gasoline) that keeps abuses in check. Even if two merging entities happen to have a strong retail presence in particular local markets, regulators have quick and effective remedies: they can (and usually do) order the firms to sell gas stations and refineries.

There is also intense competition upstream, in the wild world of oil and gas exploration. Petroleum is probably the only global business in which the industry's largest firms and best assets are controlled by governments. It may seem astonishing, but even the likes of Chevron and Texaco are midgets compared with the state-run oil giants like Saudi Arabia's Aramco. The industry's lowest-cost reserves are also controlled by governments (think Saudi Arabia and

Iraq, for a start). This leaves the private sector to fight ferociously over those oil and gas fields still left to be discovered in remote parts of the world; even here, they are finding that once-sleepy state firms from Brazil, China, and elsewhere are now partially privatized and competing with them for exploration rights.

Long gone is the heyday of the scrappy independent oil producer striking black gold in its backyard. The action nowadays is in such places as the deep waters off West Africa and Brazil, where the risks are high and the capital costs of exploration and production enormous—as much as $6 billion a pop—as are the potential rewards if a huge oil field is discovered. Only well-capitalized, large firms can ante up in this poker game and hope to thrive. Being bigger clearly helps oilmen desperate to meet the demands of institutional shareholders for sustainably high returns.

The discovery of oil in the North Sea played an important role in checking the excesses of OPEC's power. Those fields are now getting old and, in time, will yield less oil. The rise of super-majors increases the odds that other sources of non-OPEC oil will be found, which will help keep the cartel's power in check. Though it does not make for a good stump speech on the road to the White House, that is surely good news for consumers.

The big firms are even expanding into nonconventional oil resources like tar sands and heavy crude oil. That will pose environmental problems if global warming turns out to be as serious a threat as now seems possible (see chapter 5), but it could also prove a supply-side check on OPEC. Canada, for example, has deposits of tar sands (mucky hydrocarbons that can be converted into a substitute for oil at greater expense and harm to the environment) with the energy potential of all the oil under the Saudi desert. The snag is that companies will not invest to bring much of that nonconventional oil to market unless they can be assured of reliably high oil prices—and, as the past few years have demonstrated, OPEC's specialty is producing volatile, not consistently high, prices. Even so, Sir Mark Moody-Stuart, the former head of Shell, thinks that such "nontraditional oil will eventually behave like non-OPEC oil or marginal fields do today: if OPEC raises prices too much, these sources will help regulate the price."

Drowning in Oil

If you think OPEC ministers are a conspiratorial cabal, you ought to meet the Depletion Doomsday gang. This colorful group of petroleum geologists is convinced that the world is perilously close to an oil shock induced by scarcity, not politics. Several dozen of these pessimists got together recently in a dingy auditorium at London's Imperial College for a most peculiar planning session. Leading lights of this movement, including Colin Campbell and Jean Laherrère, presented technical data on oil-depletion rates that supported their grim prognosis. Their experts ridiculed rival analyses, done by American government agencies and the International Energy Agency, that contradicted their views. Campbell even decried the "amazing display of ignorance, deliberate ignorance, denial, and obfuscation" by governments and academics on this topic.

The air was thick with talk of conspiracies. During an open discussion session, participants could hardly contain their anger. They tossed out questions like "Why are there hardly 300 experts working on oil depletion anywhere in the world, fewer even than working on godforsaken neutrinos?" "Why do governments ignore our warnings?" "Don't they know that a terrible crisis is coming if we do nothing?" The gathered heretics agreed to cement their orthodoxy by forming a new think tank—the Oil Depletion Analysis Centre—devoted to the topic of oil depletion. They closed by agreeing to "go forth and spread our message urgently."

Such insistent views cannot be dismissed out of hand. Some people in this camp—such as Princeton University's Kenneth Deffeyes, as well as Campbell and Laherrère—are experienced and respected geologists. Their message is a serious one indeed, if true. Even if the West is less vulnerable to an oil shock these days, and OPEC less inclined to produce such a shock, the world economy would still be devastated if oil reserves pass a critical threshold. So is the oil really running out? This nonrenewable resource has to run out someday, but the balance of evidence suggests that day is decades, not years, away.

Doomsters have been predicting dry wells for decades, but the oil is still gushing. Vast sums have been spent and professional reputations staked on trying to guess what proportion of the total amount of conventional oil in the ground mankind has already consumed. Extreme pessimists such as Campbell and Laherrère argue that depletion is already around the halfway mark. That threshold not only is psychologically important but also marks the inflection point past which supply cannot hope to keep pace with ever-rising demand. Most mainstream forecasters, such as the experts at the U.S. Geological Survey, argue that this turning point is still decades away. The forecasters at Shell expect the halfway mark to be reached sometime between 2015 and 2030. The International Energy Agency agrees that many oil fields outside the Middle East will soon mature, but it does not expect a global supply crunch in the next couple of decades.

Pundits rarely get their oil forecasts right. Nearly all the predictions made in the wake of the 1970s oil shocks for oil prices in 2000 were well off the mark. America's Department of Energy, for example, thought that oil at the turn of the century would top $150 a barrel (at today's prices). Planners at Exxon predicted a price of $100, which was still way out, but at least their forecast for oil demand in 2000 was spot on.

One of the best forecasting records belongs to Morris Adelman, a veteran energy economist and professor emeritus at the Massachusetts Institute of Technology. He has long insisted that oil is not only plentiful but also a "fungible, global commodity" that will find its way to markets regardless of politics, making nonsense of all the talk about energy security and independence. "Back in 1973, I predicted in *The Economist* that if the Arabs don't sell us oil, somebody else will," he recalls. One reason for his optimism has been the poor quality of information about reserves in many parts of the world. It turns out that there is much more oil hidden away under the earth's surface than most people imagined back in the 1970s, and technology is making much more of it recoverable. Exxon says it has learned one crucial lesson from earlier forecasting mistakes: it greatly underestimated the power of technology. Thanks to advances in exploration and production technology, the amount of oil squeezed out of existing reservoirs and tapped from remote new ones has increased enormously. Even hitherto uneconomic hydrocarbons, such as Canada's vast deposits of tar sands, are becoming more attractive and less expensive, thanks to technological advances.

But can this frenzied pace of innovation continue? "You must be kidding: we're just getting started," says Euan Baird, who served for many years as the boss of Schlumberger, a giant oil-services firm. Chapter 9, which surveys the breathtaking advances taking place in oil-exploration technology, explains how drilling today resembles rocket science much more than it does the hit-or-miss gushers of yesteryear.

The market itself provides reassurance that the recent oil price surge does not reflect physical scarcity. In a commodity market free of manipulation, consistently rising prices would indeed be a sign of resource depletion. Oil, of course, is not a free market thanks to OPEC's machinations. However, for a brief period back in 1998 the cartel was in such disarray that the market began pricing oil as a normal commodity; that's when prices collapsed. That interplay of supply and demand produced a price of just $10 a barrel. The Algerian oil minister of the time even suggested that the price would drop to below $5 a barrel if OPEC collapsed altogether—a sentiment echoed by Venezuela's President Hugo Chavez in late 2001 as OPEC's grip weakened once again. That price response clearly suggests that there is plenty of oil around.

But Adelman goes further. He points out that tomorrow's oil-exploration technology is bound to be better than today's. He dismisses the idea of an oil crisis in the short to medium term: "Scarcity is still assumed even by reasonable men and middle-of-the-road forecasters, but that is wrong. For the next twenty-five to fifty years, the oil available to the market is for all intents and purposes infinite."

Warning: Volatile Substance

Even if the petro-pessimists are Chicken Littles, the world's reliance on oil is still problematic. Volatility, environmental concerns, and especially the future concentration of reserves in the hands of very few countries should encourage us to start weaning ourselves off petroleum now.

Price gyrations impose real pain on the economies of both producer and consumer nations. The world was lulled into a false sense of security by the decade-long period of low and stable prices following the collapse of oil prices in the mid-1980s. Taking a longer view, however, volatility in oil prices appears to be the norm, as it is for every other commodity. Indeed, it seems worse under the fractious and ill-disciplined OPEC oil cartel than it would be either in a free market or under a strong monopoly.

What is more, changes in the oil industry in the past few years have had the effect of increasing volatility. According to economist Michael Lynch, the stable oil prices that consumers enjoyed during most of the 1990s may not return anytime soon:

> There is no question that oil price volatility has increased in the past few years, reflecting a combination of causes. Like all commodities, oil suffers from volatility due to uncertainty about supply and demand, as many of the driving factors—weather, GDP growth, and so forth—remain volatile and uncertain themselves. Additionally, after a decade and a half of high surplus capacity for crude oil production, shipping, and refining, the industry has largely returned to equilibrium, meaning that there is much less ability to increase production during periods of market tightness. Like other organizations before it, OPEC is generally seeking to stabilize the price of oil, but it faces a daunting task in the near-term due to both those uncertainties, and the poor quality of the data.

In Lynch's view, the biggest reason the oil market was steady for so long was the overhang of excess oil-production capacity in OPEC. But that came to an end as investment by cartel producers failed to keep pace with soaring demand. By 2003, only Saudi Arabia had much spare capacity left.

To be fair, OPEC is not the only culprit. The rise of market forces (in what remains a highly flawed market) has also contributed to volatility. For example, recent consolidation and tightness in the oil-storage and tanker-shipping businesses have made the market jumpier. These may act as de facto production constraints, since OPEC will struggle to get oil to market quickly even if it decides to produce lots more of it. Another factor contributing to volatility has been the move by big oil companies toward just-in-time management of stocks and deliveries. Firms are keeping far lower inventories than they might have at comparable times of the year a decade ago. This is good for shareholders, as their capital is no longer tied up in excess inventory. But it may also mean that the industry loses a valuable buffer. OPEC itself appears to have embraced such a low-inventory approach. The cartel may have been inspired by Mobil's successful cost-reduction campaign before its acquisition by Exxon, known as K.I.L.L.: Keep Inventories Low and Lean. A related complication is bottlenecks in the refining and pipelines system. No new refinery was built in America in the 1980s or 1990s despite strong growth in demand. A combination of low margins, red tape, tougher environmental regulations, and the N.I.M.B.Y. (not in my back yard) syndrome conspired to make this part of the oil business thoroughly unattractive.

Another factor that kept oil markets tight in the early years of the twenty-first century (even before Bush started talking of an Iraqi invasion) was that $30-plus oil did not produce the response seen during past upswings in the oil cycle: an immediate orgy of upstream spending by non-OPEC firms that would temper crude prices. The oil business has undergone a radical transformation over the past few years, away from a fixation on volume and market size and toward financial targets such as returns on capital. The culmination of this trend was the wave of mergers that swept through the industry, producing such giants as ExxonMobil and BP Amoco (now BP). As chapter 2 explained, the new mantra of Big Oil is financial prudence.

The most powerful force contributing to oil's volatility, as Europe's gasoline riots demonstrated, is the black stuff's paramount importance in transport. During earlier shocks, developed economies were grossly inefficient in their use of oil; since then, governments have used such tools as energy taxes to make their economies less reliant on oil. They have largely succeeded, except in transport—where, despite soaring gasoline taxes, oil will always remain king while the alternatives are expensive and impractical. Most OPEC oil now goes to a sector that cannot at present live without it.

Lynch offers a blunt assessment: "Oil market volatility cannot be eliminated short of massive government intervention at the international level, and the history of commodity price stabilization suggests that this would be phenomenally expensive, if not fruitless." The upshot may be that, even after all the lessons learned by rich countries from earlier shocks, oil still has the ability to humiliate Western leaders and batter their economies.

Getting Greener

Most oil bosses will tell you that their biggest worry for the future is not oil scarcity but the environment—or at least environmental politics. At the global level, their main headache is climate change. For years the industry has tried to ridicule the environmental brigade and dismiss any scientific evidence suggesting that burning fossil fuels might contribute to global warming. It has been forced to soften its stance. The clearest sign of change is the progress, albeit in fits and starts, of the Kyoto Protocol, a pact among industrial countries to cut emissions of greenhouse gases.

Whatever happens with the Kyoto treaty, many energy firms

now accept that national restraints on carbon emissions are likely to be introduced in the medium term. Indeed, a growing band of companies, led by BP and Shell, are already preparing for the day when the price of carbon emissions is no longer zero. Shell's former boss Sir Mark thinks the Kyoto pact is crucial because it forces businesses to "put their best and sharpest minds on the task" of reducing carbon emissions. Other firms, notably Exxon, scoff publicly about global warming but are actually busy investing huge sums in energy efficiency, geological "sequestration" (a nifty way of stripping the carbon out of hydrocarbon fuels and storing it underground), and other low-carbon technologies as a way to hedge their bets.

If oilmen are getting headaches over global warming, they are suffering migraines over local air pollution. Concerns about the environmental and human costs of burning fossil fuels have risen to the top of the agenda in many countries. The rich world is imposing ever-stricter emissions standards on refineries and power-generation plants, as well as tightening the requirements to reduce pollutants in gasoline. And poorer countries too, as they gradually become better off, are putting pressure on energy companies to clean up. "Concern over the environment will not be linear, but it will be an extremely significant and irreversible force over the longer term, especially in developing countries," says the boss of an oil giant. That explains why Shell and BP have made a big shift toward cleaner natural gas and have placed smaller but still significant hedging bets on renewable energy and hydrogen.

Taken together, worries about global warming and local air pollution (described in turn in the next two chapters, on the environment) pose a serious threat to oil's future. That alone may spur a move to a post-petroleum world. Even if it does not, there is one other factor that might: the consumers of the world will grow ever more dependent on the willingness of OPEC countries—and especially Saudi Arabia and Iraq—to supply oil from their enormous reserves.

Concentrated Power

Political leaders with foresight would see that genuine energy independence comes not from adding a spoonful of Alaskan oil or a dollop of conservation (never mind just leaving things to the distorted workings of the oil market), but from encouraging the speedy development of alternatives to oil-fired transport. Thanks to rapid developments in fuel cells and hydrogen technology, there is reason for optimism that the world may yet move beyond petroleum.

It may turn out that critics of Ali Naimi, the Saudi oil minister, have the last laugh in the end: they are probably right in predicting oil's demise, even if they are wrong about the causes. Still, it might be a long wait. As the old hands of the energy business like to say, "The best substitute for gasoline is gasoline," and it seems pretty plain that oil will be dislodged only by something that is equally cheap, easy to use, and efficient—and less polluting. Even if such a thing can be found, oil could be around for decades yet, so great are the sunk investments in infrastructure, so strong the power of incumbency, so impressive the advances in fossil-fuel technology—and, quite possibly, so vast the remaining deposits.

If still in doubt, visit Al-Shaybah. Flying over the inhospitable expanses of Saudi Arabia's Empty Quarter, you will see nothing but a desolate stretch of desert, larger than France; yet tucked away under the striking red sand dunes is one of the world's largest oil fields. Aramco, the government oil company controlled by Naimi, had to invest a total of $2.5 billion to bring the Shaybah field onstream, but since it opened in 1999, its vast output of some 600,000 barrels a day has already more than repaid that investment. For decades to come, everything it produces will be undiluted profit. And as you stare down on Shaybah, consider that Saudi Arabia could develop another dozen fields of the same size without beginning to make a dent in its proven reserves. Oil may not be tomorrow's fuel, but unless the world's leaders show more foresight than they have so far, today could go on for an awfully long time.

ENVIRONMENTAL PRESSURES

The Green Dilemma

5

Welcome to Global Weirding

SOME CONSIDER the Maldives a paradise. The country is made up of more than a thousand coral islands strewn carelessly across the turquoise waters of the equatorial Indian Ocean. Buoyed by perfect weather, prime location, and (until now) positive karma, the country has become a favorite of sun worshipers and scuba divers the world over. The tourism boom has given the country's quarter million or so locals relatively comfortable lives, at least by the miserable standards of south Asia. How much longer the Maldives will remain such a paradise is unclear. The sea around these low-lying spits of land is rising, and the alarmed residents have no doubt about the culprit: global warming caused by the rich world's guzzling of fossil fuels.

The problem of greenhouse gases may sound nebulous in Manhattan or Munich, but to Maldivians it is an all too present danger. The noisy tea shops of Male, the country's capital, are full of men (but not women, as it is a traditional Islamic society) who can talk for hours about the early signs of impending disaster. Fishermen at the crowded and smelly market complain of the scarcity of live bait. Hotel owners lament that the arrival of warmer waters has bleached the colors out of much of the coral, and they worry that future tourism will suffer.

Maumoon Abdul Gayoom, the president of the Maldives and Asia's longest-serving ruler, uses his iron grip on the local press not only to keep himself and his circle of friends in power but also to keep the dangers of climate change close to the front of the national debate. In *The Maldives: A Nation in Peril*, he explained the traumatic episode that converted him to this cause:

> The impact of an environmental disaster was made appallingly real for me one day in April 1987. Unusually high waves struck Male and other islands in the Maldives with a ferocity that inflicted enormous damage throughout the country . . . While I was inspecting the damage, a large wave reared up suddenly and buffeted the vehicle I was in. It dragged me and my colleagues towards the open sea. It was a moment of fear, not for my own safety, but for the safety of the people of the Maldives, and the future of our country.

Since then, he has become a passionate international spokesman for low-lying islanders everywhere. Arguing that poor island-states like his are innocent victims of the excesses of the industrial world, he insists that rich countries must change their ways fast—and help islanders to adapt: "The Maldives believes in the polluter pays principle . . . small island-states do not have the financial means or the technological know-how to effectively deal with these problems."

The trouble is that most of the Maldives land is less than three feet above sea level. If the oceans rise just two feet or so this century, which many who study climate are predicting, the country would be destroyed. Even an increase of just a foot, combined with the bigger waves that are expected, could wreak havoc. William Allison, a scientist living in Male, explained that global warming threatens local coral in several ways. Rising seas are not bad in themselves, he said, as they give coral more upward growing room. But coral flourishes in water of around 22°C. A warmer atmosphere would heat the local ocean beyond that temperature, killing the coral. Another danger comes from carbon dioxide. When too much CO_2 dissolves in seawater, he explained, corals build skeletons only with the greatest difficulty. His final fear—as yet unconfirmed—was that hotter temperatures could dramatically increase the scale and frequency of storms.

To combat the surge in dangerous waves and storms, the Maldives government built a six-foot-high concrete barrier that partly rings the capital. Local wags call it the Great Wall of Male. Set just offshore, it was designed to absorb wave energy and spare Male further damage. More ambitious were plans for Hulhumale, an artificial island higher than Male that officials said could one day house up to half the country's present population. Another government plan proposed transferring people from the smaller islands to three bigger ones, which would be defended behind seawalls. Such decisions are not made lightly. The country's environment minister talked fondly of small-island life, particularly its sense of community, which he felt was missing in the bustle of Male. The outliers may not have a choice: "We simply cannot have 200 inhabited islands, one with 60,000 people and others with 200, vying for the same expensive defenses and services!"

But seawalls have drawbacks of their own. The coral that forms the islands is porous, making the islands in effect giant sponges. If the ocean continues to rise, before long the salt water will begin to seep through from under the seawalls. In any case, costs would be staggering: the Great Wall of Male cost an estimated $4,000 per foot to build. An official explained that the Japanese government was generous enough to pay for it. He then hesitated, biting his tongue. Yes? He finally blurted out that the aid was linked to a building contract for a foreign company, which used patented technology. To extend or repair the wall, he groused, they were forced to buy from the firm at outrageous prices. "These rich countries pollute the atmosphere," he said as anger displaced his jolly smile, "and then they profit from it."

This tale seems heart-wrenchingly clear: peaceful denizens of a simple land, in harmony with their environment, paying for the profligacy and pollution of others. That this culture might soon be lost under the rising seas like the glories of Atlantis seems a moral outrage that demands immediate and decisive action.

But the Maldives story is not so simple. Although global warming does pose a threat to coral, the severe bleaching that wiped out much of the country's coral in 1997 and 1998 was actually the result of an unusually strong appearance of El Niño, the cyclical weather phenomenon that has buffeted these islands for decades. Joan Kleypas, an expert at America's National Center for Atmospheric Research, acknowledges that the bleaching that season wiped out perhaps 15 percent of the world's reefs—but, she insists, "I won't go out on a limb and say El Niño is related to global warming."

The sea level is indeed rising, and a warming atmosphere is contributing to that trend. However, that may not be the chief reason why locals are now huddling behind seawalls in Male: the Great Wall was probably needed, say local experts who dare not speak publicly, only because the island's natural wave buffer—its wide, flat reef—was filled in to house a booming population. Local historians recall a time when the Maldivians were more nomadic. Life was not always so idyllic, and the climate not always calm. In the past, when storms destroyed one island habitat, they would move to another. Nowadays, thanks to the arrival of modern amenities like climate-controlled concrete buildings and asphalt roads, they would rather stay put if they can.

A lot of the finger-pointing directed at the rich world has a tinge of hypocrisy to it. Although the big industrial countries are without question the main source of man-made greenhouse gases, the locals are not particularly green either. Male's residents made such a call on the underlying aquifer that the groundwater got laced with salt.

Their freshwater (and its funny-tasting Coca Cola, for that matter) reaches them by way of desalinization plants. Even fresh air is getting scarce. Absurd as it may seem, the tiny city of Male has terrible traffic jams. Locals take their cars to travel the distance that people in most other countries would not bother taking a bicycle. Many idle their engines even when standing still, just to run the air-conditioning.

In short, keeping paradise afloat is not the only problem confronting Maldivians: they need to figure out how to keep paradise paradise.

As this tale from the tropics suggests, no aspect of the global warming saga is simple. The scientific evidence is controversial, with some experts claiming that mankind is clearly at fault, while others insist that there is no warming going on at all. The likely impact of any such warming is also hotly contested, with some arguing that a hotter earth would be better for mankind. The debate over the economic impacts of warming, and over efforts to slow or stop that warming, is even more polarized: many greens and techno-optimists claim that the problem can be solved at little cost, while much of the fossil-fuel industry argues that the costs could devastate the global economy. Predictably enough, the politics involved is colorful, rancorous, and too important to ignore.

To make sense of all the confusing claims and counterclaims, a few basic questions must be answered: Is global warming really happening? Why should we care? What are the possible responses? How much will responding cost? And who pays?

There are few straightforward answers to these questions, but tackling them honestly leads to an inescapable conclusion: in the long run, we must shift to a radically different energy system that releases little or no greenhouse gases into the atmosphere. Such a change could very well lead to the green nirvana of renewable energy. However, that does not necessarily mean a world powered only by windmills and bicycles. It is possible that the green nightmare of nuclear power could benefit, as it also happens to be free of greenhouse gases (though chapter 10 will argue that such a renaissance is unlikely). Indeed, even coal, the filthiest of fossil fuels, could play a big role in that climate-friendly future if technologies for using it cleanly are perfected.

This energy revolution can start slowly, but it has to start soon if the worst potential impacts of climate change are to be avoided. And there lies the rub: the enormously powerful vested interests maintaining the status quo are making it difficult for the world to take even tiny steps toward the clean energy future.

Churchill and the Climate Contrarians

What would a leader like Winston Churchill have done about climate change? Imagine that he had been presented with an emerging problem that could, if neglected, turn into a global disaster. Imagine that a response might require concerted global action and perhaps even economic sacrifice on the home front. Now imagine that his aides could not provide him with irrefutable evidence of that impending crisis. Would he have done nothing—or would he have started taking sensible precautions despite the uncertainty?

The answer is obvious, you might think. After all, history teaches us that Churchill did not dismiss the Nazi threat for lack of conclusive evidence of Hitler's evil intentions. He prepared a deliberate plan of attack, organized a global alliance, and girded his citizenry at home for a long fight. In other words, he showed leadership. Ah, but now imagine that Churchill's advisers had added the following provisos: the evidence of this problem will remain cloudy for decades, and the worst effects may not be felt for a century, but the costs of tackling the problem will probably start biting immediately. Would the great man really have rushed to act in 1940 to prevent potential catastrophe in 2040?

That, in a nutshell, captures the dilemma of climate change. It is asking a great deal of politicians to think, let alone act, on behalf of voters who have not yet been born. The problem is that as society grows ever more sophisticated in its use of science and technology, mankind's ability to influence the natural environment—and its ability to measure and analyze that impact—grows along with it. That means we are likely to be confronted in coming decades with other global science scares that promise their own irreversible "triggers" (or points of no return), and we will need to learn how to distinguish which are genuine and which bogus.

In the wake of the recent political rows over the Kyoto Protocol—the UN climate treaty George W. Bush pulled America out of in 2001 amid much acrimony—it is clear that we need to think hard about how to make decisions sensibly in an uncertain world. And when it comes to climate, uncertainty looks to be with us for a very, very long time. Richard Lindzen, an MIT meteorologist, is one of the noisiest of a noisy band of climate skeptics. On the opinion page of *The Wall Street Journal*, which often argues that global warming is nonsense, he wrote:

> The climate is always changing; change is the norm. Two centuries ago, much of the Northern Hemisphere was emerging from a little ice age. A millennium ago, during the Middle Ages, the same region was in a warm period. Thirty years ago, we were concerned with global cooling.
>
> Distinguishing the small recent changes in global mean temperature from the natural variability, which is unknown, is not a trivial task. All attempts so far make the assumption that existing computer climate models simulate natural variability, but I doubt that anyone really believes this assumption.
>
> We simply do not know what relation, if any, exists between global climate changes and water vapor, clouds, storms, hurricanes, and other factors, including regional climate changes, which are generally much larger than global changes and not correlated with them. Nor do we know how to predict changes in greenhouse gases. This is because we cannot forecast economic and technological change over the next century, and also because there are

> many man-made substances whose properties and levels are not well known, but which could be comparable in importance to carbon dioxide.

Many climate skeptics go much further and trash the notion of global warming altogether, but usually their arguments are based more on polemic than peer-reviewed science. In contrast, the pugnacious Lindzen can usually back up most of his provocative assertions. He even makes the clever argument that he is not a climate skeptic at all: "Although there is certainly room for skepticism, scientists who note the profound disconnect between the scientific meaning of common statements and the public interpretation are not being skeptical. They are nonetheless designated as skeptics in order to marginalize their views."

What may come as a surprise is that most mainstream climate scientists, who agree that global warming is really happening, would readily concede Lindzen's assertion that many uncertainties remain—though these caveats usually do not make the headlines. Everyone agrees, for example, that one big unknown is the relationship between oceans and climate. The oceans, which absorb carbon from the atmosphere as part of nature's "carbon cycle," act as a time-delay mechanism. Their massive thermal inertia means that the climate system responds only very slowly to changes in the composition of the atmosphere. Another complication arises from the relationship between carbon dioxide, the principal greenhouse gas, and sulfur dioxide (SO_2), a common pollutant. Efforts to reduce man-made emissions of greenhouse gases by cutting down on fossil-fuel use will reduce emissions of both gases. The reduction in CO_2 will slow warming, but the concurrent SO_2 reduction may mask that by contributing to a slight warming effect. There are so many similarly fuzzy factors—ranging from aerosol particles to clouds to cosmic radiation—that many parts of the world could endure unfamiliar weather patterns and maybe even freakish storms for years without knowing how it is happening or what to do about it.

There is a strong temptation to blame global warming for

perceived weather oddities. The summer of 1995, for example, brought an unprecedented and unrelenting combination of heat and humidity to Chicago and its environs that left more than 700 people dead; exactly one year after that tragedy, the same region was overwhelmed by its worst flooding ever recorded. Many blamed global warming. In 2000, uncharacteristically powerful storms and floods in Mozambique caused an international humanitarian crisis when many villagers fled their homes. That same year, parts of Britain were also deluged by monsoon-like rains that led to severe floods that greatly traumatized the nation. *The Ecologist*, a British magazine, captured the panic well with a special edition that featured a photo of three desperate men racing for their lives ahead of raging floods that had already overturned a house in the background. The headline screamed CLIMATE CHANGE: IT'S FASTER THAN YOU THINK. In 2002, parts of Central Europe were devastated by heavy rains that flooded such historical gems as Dresden and Prague. Jürgen Trittin, the German environment minister, cried out in anguish that this storm of the century was the legacy of a hundred years of reckless industrialization.

How else to explain the series of bizarre and often deadly climate catastrophes of recent years, right? Wrong. As any good scientist, no matter how firm a believer in global warming, will tell you, no single weather event can be linked conclusively to climate change. Taken together, such events are consistent with the sorts of things a warming trend might bring—but we cannot be sure. In fact, many of these seemingly extraordinary events might well fall within the natural variability of the earth's climate. That prudent stance will not satisfy those people in Peoria or Prague who want to know why their tulips are turning up earlier each year. Perhaps we should call this trend global "weirding" until the picture clears up.

But that day of scientific certainty looks to be far in the future. Tom Wigley explains why. As a leading member of the UN's Intergovernmental Panel on Climate Change (IPCC), he is at the heart of the global scientific consensus that climate change is genuine enough to take seriously. However, he also argues that whatever policy changes governments pursue in response to climate change (including doing nothing), scientific uncertainties will conspire to "make it difficult to detect the effects of such changes, probably for many decades." Yes, *decades*.

As evidence, he points to the difference in temperature resulting from an array of emissions "pathways" on which the world could choose to embark if it decided to tackle climate change. He plots various strategies for the reduction of carbon dioxide—including the one associated with the UN's Kyoto treaty—that would lead in the next century to the stabilization of atmospheric emissions of CO_2 at 550 parts per million. That is roughly double the level that prevailed in preindustrial times, and it is cited by many climate scientists, policymakers, and environmentalists as a reasonable target. However, Wigley concludes that even in 2040, the temperature differences between the various trajectories (each representing a different policy choice) are so tiny as to be nearly negligible—and certainly within the magnitude of natural climatic variance. In other words, we probably will not know if we have overreacted or not done enough to avert disaster.

As that notion sinks in, it is tempting just to throw our hands in the air and do nothing. After all, how can we possibly achieve anything if we are likely to be flying blind for decades? Yet there are some points of light to help us through the haze obscuring the linkages between mankind's use of energy and the earth's atmosphere. Though most climate skeptics will not advertise it, the question marks still hanging over climate science do not mean we know nothing.

The Goldilocks Planet

For starters, we know that *the "greenhouse effect" is real.* Without the heat-trapping effect of water vapor, CO_2, methane, and other naturally occurring greenhouse gases, our planet would be a lifeless 30°C or so colder. That would make it more like the atmosphere on Mars. On the other hand, too much of this effect is no good either. Just look at Venus, our other planetary neighbor: an intense greenhouse effect produces temperatures above the boiling point of water and makes it as hostile to life as Mars. That's why mankind needs to worry about greenhouse-gas concentrations in the first place: like Goldilocks, we need our planet not too hot and not too cold.

We also know that *concentrations of greenhouse gases in the air have been increasing*. CO_2 levels held pretty steady, around 280 parts per million, for the thousand years before 1800. Since then, as industrialization really got under way, atmospheric concentrations of CO_2 started to rise; they are around 370 parts per million today. Carbon dioxide is produced when fossil fuels are used, or when forests are denuded and burned. Agriculture and other land use releases methane and nitrous oxide, which are also powerful greenhouse gases. Industrial processes release chemicals known as halocarbons—such as chlorofluorocarbons (CFCs), chemicals used in refrigeration that have been found to deplete the ozone layer—and other long-lived gases such as sulfur hexafluoride, which are also known to contribute to climate change.

These releases matter because in the long term, the laws of nature dictate that the earth must release heat into space at the same rate at which it absorbs energy from the sun. By increasing the atmosphere's absorption of infrared radiation, such man-made greenhouse-gas emissions will force the climate to somehow restore the energy balance. One likely way is a warming of the earth's surface and lower atmosphere, but the experts expect other changes too: in cloud cover and wind patterns, for example. For reasons that we do not yet fully understand, some of these changes are likely to act as positive feedbacks that speed the warming, while others will probably do the reverse.

That leads to the next thing that we know: *the earth is warming up*. The global surface temperature has increased by about half a degree Celsius since 1975. That may not seem like much to most people, but global average temperatures (unlike the fluctuation in temperature in a particular spot during the course of a single day) do not change that much that quickly. In fact, that surge of warming has sent the global temperature to its highest level in a thousand years. America's official National Oceanic and Atmospheric Administration reckons that the last few years of the 1990s were among the "warmest on record," based on data going back over a century. Dissenting voices have refused to accept this evidence—which is drawn from ground-based observations as well as measurements made by ships at sea—as they claim that the techniques used are unreliable and inconsistent. They point to satellite data suggesting that the lower troposphere (the part of the atmosphere from ground level up to eight kilometers) has not been warming up very much. This inconsistency, they argue, casts doubt on the entire notion of global warming.

This is a serious charge, and one that the scientific establishment has taken seriously. America's National Research Council established a panel of experts to investigate. The group concluded that the skeptics have no case: the warming trend is "undoubtedly real," it said, and added that "the disparity between surface and upper air trends in no way invalidates the conclusion that surface temperature has been rising." There is another place to look for empirical evidence that the earth is warming up: the deep blue sea. The ocean is the only place where energy from what scientists call a "planetary radiation imbalance" can accumulate over the long term; the thermal conductivity of land is simply too low to absorb much of that energy. Scientists say that since the mid-1950s the energy content of the seas has indeed increased significantly by an amount roughly consistent with the observed elevation in temperatures.

John McNeill of Georgetown University sums up the significance of this warming trend in *Something New Under the Sun: An Environmental History of the Twentieth-Century World*:

> These minute but momentous changes [in CO_2 and methane concentrations] in the atmosphere, in concert

> with some even tinier ones involving other greenhouse gases, made the atmosphere more efficient at trapping heat from the sun. At the same time, human actions injected lots of dust and soot into the atmosphere, which slightly lowered the amount of solar energy reaching the earth's surface. The net effect, since 1800 or so, was about 2 more watts per square meter of solar energy delivered to the earth's surface. This probably accounts for the modest warming the earth experienced in the twentieth century.
>
> The earth has warmed up recently, although no one knows for certain if human actions are the cause. Between 1890 and 1990, average surface temperatures increased by 0.3 degrees to 0.6 degrees Celsius. That happened in two surges, between 1910 and 1940, and then after 1975. From 1940 to 1975, average temperatures actually declined slightly. But nine of the ten hottest years on record occurred between 1987 and 1997, and the 1990s promised to be the hottest decade since the fourteenth century. Changes of this magnitude and rapidity are well within the natural range of variation, although rare within the last 2 million years, probably nonexistent within the last 10,000 years, and definitely absent within the last 600 years.

The earth is indeed warming up at an unusual pace. The most obvious explanation for this trend is also the most controversial part of the global warming debate: mankind's actions.

The Human Touch

More than a century ago, a farsighted scientist came up with a radical insight that now seems obvious to many people: burning fossil fuels has an effect on the atmosphere. Svante August Arrhenius, a Swede who went on to win the Nobel prize for unrelated work he did in chemistry, theorized in 1895 that the CO_2 released by burning fossil fuels would trap heat from the sun that would otherwise have been reflected back into space, the way the glass panels of a hothouse do. He even came up with elaborate mathematical models to calculate how much coal would need to be burned and how long it would take to double the atmospheric concentrations of CO_2—an incredibly prescient exercise. Curiously, though, he was not particularly worried about the warming trend that he saw coming. Perhaps influenced by the frigid Swedish winters, he envisioned "more equable and better climates." Although that latter prediction remains hugely controversial, there is mounting evidence to suggest that his basic insight was right: man's actions are helping fuel global warming.

James Hanson is a modern-day Arrhenius. A distinguished researcher at America's National Aeronautics and Space Administration, Hansen shot to prominence in 1988, when he sounded the alarm about mankind's role in global warming. When he began researching climate change as a young man, some people thought that the temperature trend was going downward; *Newsweek* had even run a cover story warning of the dangers of global cooling and the prospect for a new ice age. Hansen didn't buy it. As one of the first hard-core number crunchers in the climate arena, he applied rigorous mathematical models and high-powered computers to study the matter. His research convinced him that the general warming trend of the twentieth century, which had leveled off in the 1960s, was resuming. And when a Senate committee asked him in 1988 to testify about the matter, Hansen came up with a stunner: global warming is "already happening now," he said, and "with a high degree of confidence" we can say that mankind is playing a role in that warming.

Nevertheless, the question of human culpability remained extremely controversial among scientists. In late 1995 the world's leading climate experts were due to finalize their second big assessment report on global warming for the UN's Intergovernmental Panel on Climate Change (IPCC). They might have continued to sit on the fence were it not for some pathbreaking research done at the Lawrence Livermore national laboratory in California that found a way to recognize human fingerprints on the atmosphere. In addition to looking for greenhouse gases like CO_2, they decided also to track the history of man-made chemicals like sulfate aerosols. These particles are the inevitable by-product of industrial processes, and they act to cool the local environment by reflecting sunlight back into space.

These experts, led by a rising star named Ben Santer, produced computer models that took a stab at estimating the climate impact of both greenhouse-gas heating and sulfate cooling, yielding a result that could not plausibly be considered "natural": only human activities could have produced that combination of emissions and temperature changes. Those theoretical results largely conformed with what the record books showed had actually happened in the past few decades. After Santer presented these findings to the big shots of the IPCC, they could waffle no longer. Their Second Assessment Report, issued in 1996, concluded that "the balance of evidence suggests a discernible human influence on global climate." That drove the naysayers up a wall, and poor Ben Santer had to endure many attacks from them in coming years.

James Hansen's courageous, if exceptionally (some said unscientifically) blunt, comments back in 1988 turned him into a media celebrity and the darling of environmentalists everywhere, but he too became a target for the fossil-fuel industry. Nonetheless, Hansen produced another influential report in 2000, in which his team again took on the question of why exactly the earth is warming. Their paper analyzed all the various factors that have acted as "forcings" of the earth's climate since 1850. The authors were careful to distinguish natural forcings (such as pollution from volcanoes) from anthropogenic ones (such as the release of CO_2 resulting from fossil-fuel use), and to delineate the positive and negative impacts of each on temperature. Hansen and company concluded that "increasing greenhouse gases are estimated to be the largest forcing and to result in a net positive forcing, especially during the past few decades." The authors also noted that not enough attention had been paid in the climate debate to greenhouse gases other than carbon dioxide. By their reckoning, "climate forcing by non-CO_2 greenhouse gases is nearly equal to the net value of all known forcings for the period 1850–2000." In other words, emissions from SUVs are not the only thing to watch for: flooded rice paddies, flatulent cows, and rotting garbage also matter. Hansen's team concluded that humanity's role in climate change simply cannot be dismissed.

More worrying is the growing chorus of scientists who say that temperatures will continue rising as a result of our activities. The UN's IPCC has for years been struggling to make sense of this trend in a way that preserves the group's consensus. That has not been easy, as the IPCC's lead authors include not only true believers in climate change but also hard-core naysayers like MIT's Richard Lindzen. However, their tortuous deliberations mean that when the group issues a report, it carries enormous weight. And the report it put out in early 2001 changed the terms of the climate debate decisively.

The IPCC's Third Assessment Report came up with two striking conclusions that put all the remaining scientific uncertainty in perspective. First, the scientists stated that man's actions have "contributed substantially to the observed warming over the last 50 years." That language was considerably stronger than what the members had agreed on in the previous report five years earlier. The second conclusion was more striking. The IPCC warned that if those actions go unchecked, the warming could increase to much higher levels. The previous report had estimated the likely warming over the next century of an additional degree to 3.5 degrees Celsius; the new report changed that estimate from 1.4 degrees to 5.8 degrees Celsius over the next century.

Economists have complained, with some reason, that the statistical techniques used by the IPCC forecasters could be improved. Some have pointed out that this range—presented in full or, as alarmists like to do, by leaving out the lower end of the range ("IPCC says earth to warm by 6 degrees!")—is misleading: statistical analysis suggests that a rise of 3 degrees or less is much more likely than a 6-degree rise. Even allowing for such refinements, the message from the world's leading climate gurus could not have been clearer: our actions are very likely propelling us toward a hotter world.

Still, that was not good enough for some. Though George Bush himself publicly professed to care about global warming even after pulling America out of the Kyoto pact, many senior figures in his Administration ridiculed the whole idea. Many of his political allies (especially oil bosses like ExxonMobil's Lee Raymond, who upbraided me for being so "naïve" as to believe UN forecasts) said the IPCC's work was tainted by European greens, loony lefties, and global bureaucrats working on some un-American agenda. That explains why Bush, in May 2001, asked the National Academy of Sciences (NAS), America's most prestigious scientific body, to decide whether concern over global warming was justified or not. Notably, the panel included such prominent doubters as Richard Lindzen. Much to the disappointment of some inside the Bush White House, the group in essence agreed with the IPCC:

> Greenhouse gases are accumulating in Earth's atmosphere as a result of human activities, causing surface air temperatures and subsurface ocean temperatures to rise. Temperatures are, in fact, rising. The changes observed over the last several decades are likely mostly due to human activities, but we cannot rule out that some significant part of these changes is also a reflection of natural variability. Human-induced warming and associated sea level rises are expected to continue through the 21st century.

The Bush team could no longer credibly attack the scientific evidence of global warming once this American panel of climate experts had weighed in using such clear and pointed language.

After the NAS report, the fledgling American Administration came under renewed pressure to do something serious about climate change to prove it was not in the pocket of the energy industry. Amid much fanfare, Bush finally unveiled his answer to Kyoto on February 14, 2002—a domestic climate-change strategy designed to show that he was indeed taking the problem seriously. The plan would, he said, commit America to "an aggressive strategy to cut greenhouse gas intensity by 18 percent over the next ten years." He went on to promise that his new plan would direct the country toward "a path to slow the growth of greenhouse gas emissions and—as the science justifies—to stop and then reverse that growth."

At first blush, all this seemed to add up to a credible, non-Kyoto strategy. In fact, it was a sham: the plan was completely voluntary; it contained no CO_2 caps of any kind; and it did not even contain any "emissions-trading" schemes that market-friendly Republicans and utilities could have supported. Eileen Claussen of the Pew Center on Global Climate Change said at the time that this was "just an effort to cloak 'business as usual' in some finery. Emissions will continue to grow." Environmental groups like the Sierra Club were less refined in their prose, calling the plan "a Valentine's Day gift to corporate polluters."

Bill McKibben, a noted environmental writer, put the matter this way in *The New York Review of Books*: "Barring a deep change in public attitudes, we seem altogether too likely to drive blithely on, steering by the rear view mirror. That's what the Bush-Cheney energy plan does, and it's the reason most other nations, and UN Secretary-General Kofi Annan, reacted so undiplomatically to our national debate . . . the National Energy Policy is our real response to the IPCC, and the message couldn't be clearer: it's business as usual in the U.S."

Those seem like damning words, but the response from many defenders of the Bush policy is: So what's wrong with business as usual? Even some of those who accept that global warming is happening ask whether we really need to rush to do anything about it. After all, isn't the current energy system an essential part of the American economic miracle post–World War II? Besides, as

Richard Lindzen and others point out, the climate has always changed throughout history. Why worry now if man's actions turn up the heat by a few degrees more? Anyone who has endured harsh winters in Siberia, Syracuse, or Stockholm could be forgiven for thinking this trend just might make the planet a more agreeable place.

A Happy Hothouse—or the Road to Hell?

"Droughts will hit California and the Pacific Northwest. Sea levels are expected to rise 19 inches—which means you may need a gondola to get around Wall Street. And if you own coastal property, call your insurance broker." That alarming snippet, snatched from a local newspaper, appeared at the top of a *New Yorker* cartoon that ran in mid-2002. Under the quote, the magazine ran a full-page sketch by Michael Crawford of a businessman looking out the window of his skyscraper to find the Empire State and Chrysler buildings partially submerged. Yet there is nothing grim about the cartoon. The executive looks more amused than alarmed. And there is even a happy couple passing by his window in a Venetian gondola.

You can't read too much into a cartoon, but nonetheless this particular one did encapsulate three powerful ideas related to how the world might be affected by climate change. One is that there will be positive as well as negative impacts of climate change; indeed, the sea-level rise that wipes out the Maldives might turn part of inland New Jersey into prime beachfront property. Another notion is the paramount importance of adapting to whatever changes global warming throws our way: in other words, if your job as a dockworker is wiped out, learn how to steer a gondola. The final point is hinted at by the lighthearted tone of the cartoon: most people, especially in America, will probably not pay much attention to climate change until its effects appear on their doorstep. And even then they might still be amused.

Yet this is no laughing matter. While nobody knows what the precise impact of global warming will be, scientists do know enough to take it very seriously. The IPCC is quite blunt:

> *The stakes associated with projected changes in climate are high.* Numerous earth systems that sustain human societies are sensitive to climate and will be impacted by changes in climate. Impacts can be expected in ocean circulation; sea level; the water cycle; carbon and nutrient cycles; air quality; the productivity and structure of natural ecosystems; the productivity of agricultural, grazing, and timber lands; and the geographic distribution, behavior, abundance, and survival of plant and animal species, including vectors and hosts of human disease. Changes in these systems in response to climate change, as well as direct effects of climate on humans, would affect human welfare.

In short, we had better start paying attention. However, the climate gurus at the UN were careful to add that this will impact human welfare both "positively and negatively."

That measured assessment has opened a crack that climate skeptics are trying to exploit. If there are positive impacts, they say, why not welcome climate change? They point to the various potential benefits of a warmer world: Arctic shipping lanes, long frozen during winters, are now becoming passable all year round. A similar thawing is taking place at Siberian ports that were once useful for only a few months of each year. Think of the economic benefits to shipping and commerce! What is more, such folk argue, warmer nights and winters in temperate regions could mean money saved on heating bills and maybe even lives saved. Warmer weather in Kansas or Saskatchewan, for example, could lead to longer growing seasons and more income for farmers. To accept this scenario, of course, you must ignore the possibility that the warming trend could bring with it heat waves (of the sort that struck Chicago in the mid-1990s) or erratic patterns of rainfall and heat that harm agriculture.

A few extreme optimists argue that increasing greenhouse gases will be unambiguously good for life on earth. The Greening Earth Society, a nonprofit group set up by American electric utilities and fossil-fuel suppliers, is unabashed in its advocacy of this point of view. Its website offers this interesting question-and-answer session:

> **Q:** Is CO_2 a pollutant?
> **A:** No. CO_2 is a fundamental building block for life on earth. Life on earth is carbon based. Plants—the anchor of our planet's food chain—rely on carbon dioxide for life, itself. CO_2 is no more a pollutant than water is a poison.
> **Q:** But isn't the question, really, the concentration of CO_2 in the atmosphere?
> **A:** In some abstract sense, perhaps. The question itself assumes there is known to be some optimal concentration of carbon dioxide beyond which its effect becomes detrimental, either to life itself or by inducing catastrophic changes in earth's climate . . . there are significant scientific questions concerning the probability, timing, and magnitude of potential changes in climate from a doubling of CO_2, and greater.
> **Q:** Are you actually taking the position that carbon dioxide emissions from fossil fuel combustion are beneficial to life on earth?
> **A:** Yes. Higher atmospheric concentrations of CO_2 increase plant productivity, water use efficiency, and their resistance to a variety of environmental stresses including heat, drought, cold, pests, deficient nutrients, and air pollution.

If you believe those extraordinary assertions, each of which contains a kernel of scientific truth, then global warming really is a cause for celebration.

Don't break out the champagne yet, however. While most mainstream scientists accept that some countries will probably see positive impacts, they caution that the distribution of impacts will not be even. Also, even the same effects (say, warmer evenings) that seem positive at first glance could turn negative if the scale of warming increases. Also, climatic zones everywhere could shift toward the poles by 150 to 550 kilometers in the midlatitude regions, dragging entire ecosystems and agricultural zones with them for the ride and subjecting them to new and unfamiliar stresses. As a result, some ecosystems will decline or fragment, and those species that cannot readily adapt may die out completely.

A detailed study of climate change in Europe led by Britain's University of East Anglia said that climate change will probably be a modest boon for the northern countries, where winters might moderate and harvests improve. That would please Svante Arrhenius, to be sure. However, the researchers went on to warn that the impact could be detrimental for much of southern Europe, which may suffer from severe water shortages and crop failures, and could possibly even turn to desert. The olive groves of Spain and Greece could yet turn as parched as the dusty sands of northern Africa.

A similar analysis of North America by U.S. scientists reached broadly similar conclusions: although the economic impact on the country as a whole might be modest, regional impacts might be dramatically worse. Eileen Claussen, whose Pew Center on Global Climate Change is respected for its nonpartisan approach, sums up the threat to the United States:

> We face both increased flooding and increased drought. Extended heat waves, more powerful storms, and other extreme weather events will become more common. Rising sea level will inundate portions of Florida and Louisiana, while increased storm surges will threaten communities all along our nation's coastline. New York City could face critical water shortages as rising sea level raises the salinity of upstate aquifers and reservoirs. And a good chunk of lower Manhattan that's built on landfill could again be submerged. We can adjust to some of these things, if we're willing to pay the price. But many of the projected impacts

> are irreversible—when we lose a fragile ecosystem like the Everglades or Long Island Sound, it can never be replaced . . . In some communities, this is no longer a theoretical matter. The impacts are being felt right now. Just ask the people of Alaska, where roads are crumbling and homes are sagging as the permafrost begins to melt.

Though parts of the rich world can expect to see some benefits from climate change, it seems pretty clear that much of the poor world—which has the misfortune to live in or near the tropics—will be much harder hit. This is certainly unfair: after all, the average Bangladeshi emits barely one-hundredth the amount of greenhouse gases spewed out by his American counterpart, yet Dhaka is likely to suffer much more from climate change than the Dakotas. Robert Watson, the chief scientist of the World Bank and formerly head of the UN's IPCC, worries particularly about the impact of warming on Africa and the poorest parts of Asia. He observes with sadness and anger that people in the developing world are the least able to adapt to the coming threats to their agriculture, water resources, and other mainstays of life.

Climate change has already had at least one demonstrable effect on the environment: sea level. Partly due to man's actions, the oceans are slowly rising. A warmer earth has meant that many glaciers are rapidly melting. The biggest impact is in the polar regions, which are bellwethers of climate change. The IPCC warns that "climate change in the polar region is expected to be among the greatest of any region on earth. Twentieth century data [for the Arctic] show a warming trend of as much as 5 degrees Celsius over extensive land areas, while precipitation has increased . . . In the Antarctic, a marked warming trend is evident in the Antarctic Peninsula, with spectacular loss of ice shelves . . . The Arctic is extremely vulnerable to climate change, and major physical, ecological, and economic impacts are expected to appear rapidly." The meltdown in the north may not raise sea levels, since most Arctic ice (except that which sits atop Greenland) already floats in the ocean. The really big worry is that the huge quantities of ice in Antarctica are on land and will raise the sea level in the unlikely event that they too melt.

The impact of this ice melt, combined with the much more significant fact that the oceans expand as they warm, means that the sea level could rise somewhere between 10 and 90 centimeters by 2100. A large swath of humanity, whether huddled into megacities like London and Mumbai (as Bombay is called these days) or in low-lying countries like the Maldives and Bangladesh, is vulnerable to even a small rise in the world's oceans. What is more, rising seas could do serious damage even before the sea level increases dramatically. Experts from Germany's University of Bremen argued in the scientific journal *Nature* that rougher seas are likely to be a consequence of climate change. Increased wave action and heightened storm activity would wear down coastal defenses and increase the menace of flooding.

Don't Pull the Trigger

There is one more compelling reason for the rich world to address climate change now, despite the uncertainty: the possibility of unpleasant, unpredictable surprises. In particular, experts envision "major alterations in ocean circulation, cloud formation or storms; and unpredicted biological consequences of these physical climate changes, such as massive dislocations of species or pest outbreaks." Geologists have plenty of evidence that many ice ages in the past have been punctuated by warm periods; the triggers and transitions seem to be extremely rapid.

Man-made warming could prompt a dramatic step-function response. Consider, for example, an ocean circulation system in the mid-Atlantic that gives northwestern Europe its relatively mild climate. Scientists worry that rising temperatures just possibly might cause the collapse of this "conveyer belt," which could result in dramatic weather disruptions and much harsher winters on both sides of the northern Atlantic.

That points to the biggest fear of all: warming may cross some as yet unknown trigger and so lead to irreversible changes that transform the earth into an unpleasant or possibly unlivable environment. Remote as that scenario may be, it is not comforting to know that any attempts to stabilize atmospheric concentrations of greenhouse gases (at whatever level) will take a very long time indeed. The ocean's thermal inertia, explains the IPCC's Tom Wigley, means that it will take decades or even centuries for the climate to stabilize after atmospheric concentrations of greenhouse gases have leveled off. And even then the sea level will continue to rise in the future for centuries or even millennia. Given the potential scale and difficulty of this challenge, it is important to remember the following: mankind's contribution to warming is the only factor that we directly control; we are unlikely to persuade the sun to change its radiation patterns to suit our climate concerns.

It is time to start crafting a century-long strategy for dealing with climate change. The creation of the Kyoto Protocol in 1997 was an attempt to do just that. The UN treaty imposes mandatory targets on the rich countries of the world to reduce greenhouse-gas emissions by a certain percentage below 1990 levels. America agreed to cut its emissions by 7 percent by the end of this decade, while Europe accepted a target of 8 percent and Japan 6 percent. Under the treaty, poor countries have no obligation to cut anything, at least during the first decade, but they could benefit from schemes that allow rich countries to get credit for making certain sorts of clean energy investments in poor countries. In coming decades, many negotiators expect that at least the big developing countries like China, India, and Brazil will accept some sort of emissions targets.

Conceptually, this approach is robust, and it promises to be the basis for an enduring global regime for climate control. Some experts liken it to the successful GATT and WTO approach to trade liberalization, while others see parallels with the frameworks for negotiating nuclear arms reductions.

What a pity, then, that the Kyoto treaty is in trouble. The reason is that the world's political leaders have, by and large, lacked the

Churchillian vision necessary to tackle climate change seriously. In fact, the politics of the Kyoto process have proved so squalid, selfish, and shortsighted that in diplomatic circles, Kyoto has become a four-letter word.

Who Shot Kyoto?

PRESIDENT GEORGE W. BUSH, POLLUTER OF THE FREE WORLD. So screamed a banner headline, accompanied by a particularly unflattering picture of America's President, that ran in early 2001 in a British newspaper. And this could not be dismissed as just the latest publicity stunt pulled by the country's infamous tabloids: it appeared in *The Independent*, one of the quality broadsheets.

There are many reasons for such animosity, but first among them is Bush's truculent stance on global warming. Though he has always insisted he is genuinely concerned about climate change, his actions suggest otherwise. Soon after taking office, he caved in to pressure from several powerful senators and conservative lobbying and junked his campaign pledge to regulate emissions of carbon dioxide. Environmental groups were outraged. Many found that the media attention helped galvanize armchair greens, who started turning up at protests and sending in checks to groups like the Sierra Club in numbers not seen since Newt Gingrich's unsuccessful attack on America's environmental regulations nearly a decade earlier.

The irony is that Bush's U-turn caught even some of his own supporters in the energy business off guard. A few big coal-fired utilities had actually asked the government for CO_2 regulations so that they could invest in new power plants (which typically last for decades) with the assurance that some future administration would not suddenly impose radically different restrictions on CO_2—and so force them to make hugely expensive retrofits or to abandon plants altogether.

One such firm was quickly put in its place when it informed the

White House's energy task force that a number of power companies were ready to go further than the Bush Administration on CO_2. A company official present at that meeting, who insisted on anonymity, described it this way: "When we told them our position, they let us know very clearly that the President was not going to support CO_2 caps and that he would expect his friends in the business to fall in line with that. End of conversation. I know a shakedown when I see one, and that was a shakedown!"

After alienating many Americans, Bush did not waste time before he took on the world. Soon after his U-turn on domestic CO_2 caps, he made a noisy declaration that he would never support the Kyoto Protocol. To be fair, this was not really a U-turn: he had long insisted that he did not like the treaty or its call for binding cuts in greenhouse-gas emissions by industrialized countries. However, critics were quick to point out that his announcement seemed designed not merely to remind the world of his position but to wreck the treaty altogether. It was that take-no-prisoners approach that prompted the outraged headline in *The Independent* and other papers around the world.

Such indignation seems understandable. Bush's cursory dismissal of the Kyoto pact came after a decade of exhausting negotiations. Bush's own father had helped the process of international climate negotiations get going back in 1992 at the Earth Summit in Rio de Janeiro, while Bill Clinton and Al Gore had committed America to the final deal in Kyoto, Japan, in 1997. It did not help matters that both Bush and his powerful Vice President, Dick Cheney, have close links with the fossil-fuel lobby, which has long opposed any action on this issue. The national energy strategy they presented later in 2001 only stoked such suspicions as well: it called for hundreds of new power plants fired by fossil fuels, but it said little of substance about renewables, energy tax reform, or conservation.

Bush, however, was not the only villain. Europe's leaders deserve a share of the blame for the failure of the Kyoto negotiations. In November 2000, before Bush got to the White House, ministers from around the world had gathered in the Hague, a breezy coastal city in the Netherlands, to hash out the finer points of the big-picture agreement reached in Kyoto in 1997. The Americans, then represented by the Clinton Administration, were keen to sign. However, they were in a bind: the roaring American economy of the 1990s had set greenhouse-gas emissions soaring, and so had raised the cost for America to meet its Kyoto targets. As a result, the Americans insisted on maximum flexibility in reaching those targets through use of mechanisms such as emissions trading, storing carbon in forest sinks (trees absorb CO_2 from the air as they grow), and so on. Frank Loy, who headed the Clinton team at the Hague, argued that it does not matter to the atmosphere whether reductions in greenhouse gases are made by shutting coal plants in Ohio, funding cleaner energy in the Ukraine, or planting tropical forests in Bolivia: this is a complex global problem, he insisted, and it deserves a flexible global solution.

Though the Europeans had agreed in principle back in Kyoto to accept such flexibility mechanisms, they hardened their position after being pressured by European green groups. Egged on not to cave in to the Americans, the Europeans decided to take the moral high ground in the Netherlands. Claiming that such concessions would allow bogus cuts that would "let America off the hook," they demanded sharp limits on the domestic use of energy instead. Some Europeans also had genuine worries—for example, there is scientific uncertainty about how reliably forest sinks will absorb carbon dioxide out of the atmosphere over long periods of time—but mostly the objections were political.

Many ministers used the opportunity to grandstand. Some, like Dominique Voynet, then France's environment minister, even made it clear that they wanted Americans to feel real pain. Why? Some members of the European entourage were from green parties, which held a radical view on the matter to start with; others openly relied on experts from environmental lobbyists to advise them during the frenzied late-night negotiating sessions that characterized the Hague gathering. During a speech by Frank Loy, a protester lobbed a cream pie that hit him right in the face. Loy kept his composure and, after wiping his face, finished his remarks; the audience even applauded. It was pretty clear from the smirks and smiles in the convention center that day, however, that many people present had more sympathy for the pie thrower than for Loy.

That made Jan Pronk's job impossible. The Dutch minister, who served as the chairman of the summit, tried every diplomatic trick in the book to broker a compromise. The toughest nuts to crack were not the Americans, who were keen to strike a deal before being replaced by a new Bush team, but Pronk's fellow Europeans. A year and a half later, on the eve of the big Earth Summit II in Johannesburg, Pronk explained his dilemma: "Many people argue that we should forge ahead without the United States. I refuse . . . I always push to keep the U.S. on board, because I say don't let them off the hook." Alas, his pleas fell on deaf ears. The summit came to an ignominious and unsuccessful close as the Europeans refused to make concessions to American proposals for flexibility.

European posturing turned out to be hypocritical as well: the European Commission itself has since accepted that it will be virtually impossible for the European Union (EU) to meet its own Kyoto targets without relying heavily on the very same mechanisms they wanted to deny the United States. One keen observer of the scene defends the EU, arguing that it "wanted to limit their use largely because the U.S. planned to do nothing else other than use the mechanisms (i.e., get all the reductions through sequestration and trading)" and make no domestic cuts in emissions whatsoever. Maybe, but it is not clear who gets the last laugh in this sorry tale. The European negotiators, meeting six months later in Germany to continue those failed talks at the Hague, ended up allowing American-style flexibility in the final Kyoto treaty—without, of course, the Americans on board.

Bush was a conduit for a sense among many Americans, from the boardrooms to the barrooms, who feel that there is some sort of international conspiracy to do in their cherished lifestyle. Some such sentiments are the cynical ploys of business interests threatened by the prospect of a low-carbon world. However, many people who feel this way are decent, honest folk without oil empires to defend. Big cars, the open road, and cheap gasoline are, right or wrong, inextricably linked to the American dream. That explains why the elder George Bush, who was not nearly as close to the energy industry as is his son, told the Rio Summit back in 1992 that he was willing to act on climate change but that "the American way of life is not up for negotiation." Greens fumed, but most ordinary Americans probably agreed with him.

There is another—and highly controversial—idea about why the Kyoto treaty got into trouble. David Victor, a professor at Stanford University, insists that the biggest factor in the demise of the Kyoto treaty is—ready?—the Kyoto treaty itself. In *The Collapse of the Kyoto Protocol and the Struggle to Slow Global Warming*, a book that was published just as the treaty was melting down in early 2001, he argues that Kyoto has many flaws, but he points in particular to its reliance on international emissions trading as fatal:

> The problem with trading is that it requires allocating permits that are worth hundreds of billions of dollars. In the past, countries have been able to allocate and launch trading systems within their own borders. For example, the United States has created a countrywide emissions trading system for sulfur dioxide, a key precursor to acid rain. In Europe, governments are auctioning tens of billions of dollars of licenses for the third generation of mobile telephones (the so-called "3G auctions" or "spectrum auctions"). However, success in these limited domestic experiments offers little assurance that international permit trading will work. Trading of carbon permits across borders rests on international law, which is a weak force. Nations can withdraw if their allocation proves inconvenient, and there are few strong penalties available under international law that can keep them from defecting. Yet the integrity of an emission trading system requires the impossible: that major players not withdraw.

In other words, a crude calculation of national self-interest suggests that the Kyoto treaty would not have worked anyway, Bush or no Bush. That point was reinforced in 2003 in *Environment and Statecraft*, a sharp book by Scott Barrett, a professor at Johns Hopkins University.

Not so fast, rebuts Michael Grubb, a leading climate-policy expert who is affiliated with London's Imperial College. He rejects such skepticism about international emissions trading, arguing that those mechanisms "are intended to act like elastic, pulling countries towards their emission targets through market-based incentives." In "Keeping Kyoto: A Study of Approaches to Maintaining the Kyoto Protocol on Climate Change," which appeared around the same time as Victor's book, he and several colleagues put forth this robust defense of the troubled treaty:

> The Kyoto Protocol is rooted in ten years of global negotiations . . . [Keeping Kyoto] would demonstrate the practicability of the market-based mechanisms and yield experience with their implementation. It would demonstrate to the US that other countries are serious about tackling climate change. It would give greater certainty for private sector planning and help to stimulate the development and deployment of low-emitting technologies, and associated industries and institutions, that are required for cost-effective solutions.
>
> US insistence to look at longer term timeframes, and other proposals for longer term emission goals, can be accommodated within the framework of Kyoto's sequential emission commitment periods and are not incompatible with the Kyoto Protocol. The Kyoto Protocol can enter into force without US participation and the Kyoto first-period targets can be kept. Whether to maintain the Kyoto Protocol as signed can thus be considered independently of various proposals to negotiate targets for later time periods to suit US or other concerns, and maintaining the current

> agreement would help to secure the foundations for such negotiations.

In other words, Bush or no Bush, the Kyoto Protocol is the only game in town.

Al Gore: Hero to Zero

Perhaps the most compelling reason not to hold the younger Bush solely responsible for Kyoto's woes is that Bill Clinton and Al Gore deserve blame too. That may come as a surprise. After all, Clinton's V.P. made his passionate green beliefs public in his book *Earth in the Balance*. In it, Gore went so far as to proclaim that global warming was "perhaps the greatest danger this country has ever faced." And Gore's eleventh-hour arrival at the original Kyoto climate summit in 1997, where the talks were on the brink of collapse, was widely credited with saving the treaty.

Gore did not deserve that credit. In fact, the deal that he brokered actually contained the seeds of Kyoto's destruction three years later. For one thing, the Clinton team caved in to lobbying efforts from a broad anti-Kyoto coalition that demanded that China and India also accept mandatory cuts in greenhouse-gas emissions. That was disingenuous, for everyone had agreed as far back as the Rio Summit that the rich world would act first, since it was their emissions that had caused the problem in the first place. Yet Gore insisted in Kyoto that dirt-poor countries must show "meaningful participation" or else; quite understandably, the poor world refused to act until the big polluters did. Gore's team can also be criticized for allowing too much "hot air" in Kyoto. Under the emissions trading provisions, countries that find it expensive to make cuts in greenhouse gases at home can buy credits overseas from countries that have credits to sell. A huge loophole was slipped into that good proposal when negotiators gave in to demands from Russia and Ukraine for unrealistically generous greenhouse-gas allowances (despite the fact that their economies had collapsed since the Soviet days, and now produce far fewer emissions). The creation of these so-called "hot air" credits undermined the credibility of the entire emissions trading approach, and would later backfire on the United States in the Hague.

Even setting aside those Kyoto blunders (after all, international negotiations are a bloody business), Clinton and Gore still have a pitiful record on global warming on the home front. Though Clinton insisted that he took the threat seriously, he never spent much political capital on the issue. It is true that he had to contend with a hostile Congress, but it is also true that he never pushed seriously for any domestic measures (like a cap on CO_2 emissions) during his long tenure. David Gardiner, a former Sierra Club official who headed the White House Climate Change Task Force under Clinton, has analyzed the effects of the various voluntary schemes that were put in place during the 1990s, a period when America's greenhouse-gas emissions soared. While understandably diplomatic in his language, even he concludes that "voluntary programs are extremely helpful, but of limited effectiveness in the face of a rising tide of such emissions. By themselves, they are not adequate to significantly reduce greenhouse gas emissions in the short or long term. In the next phase of climate action, the United States will need a policy that includes at least some mandatory requirements."

The upshot of all this is that the American economy that George Bush inherited from Bill Clinton had roared through the 1990s at an unprecedented pace—and, unsurprisingly given the Clinton legacy of inaction, so too did greenhouse-gas emissions. That legacy meant that had Bush embraced the Kyoto treaty, he would have signed America up to slash its emissions by a whopping quarter or more below what "business as usual" (meaning no Kyoto-type policies) would have achieved by the end of the decade. It seems pretty obvious that emissions cuts could add up to a lot of political pain, especially for a President so close to the fossil-fuel lobby. But would they necessarily add up to a lot of economic pain for Americans at large? The answer is no. Or yes. Or maybe. It all depends on whom you ask.

Costing the Earth

Messy as they are, the politics of global warming are nothing compared to the economics of the matter. When the great debates of the age are staged, economists are usually relegated to a dark and dusty corner. What good are eggheads during trench warfare? Yet when it comes to this particular war of ideas, economists find themselves on the front lines of the battle. The reason is that the climate-change debate cuts to the heart of how we use energy. Energy, of course, is the bedrock of the world economy, and that means big money, big business, and big lobbying interests are involved.

An expert from the Competitive Enterprise Institute, a right-wing think tank in Washington, offered this assessment:

> The cost of Kyoto to the U.S. alone would be about \$300 billion per year. The resulting loss of GDP over the next ten years, about 28 percent, would be nearly triple the loss to GDP experienced during the Great Depression, which saw a drop in GDP of about 10 percent.

The Global Climate Coalition, a formidable lobbying group for heavy industry, added that efforts to curb CO_2 emissions "would eliminate millions of American jobs, reduce America's ability to compete and force Americans into second-class lifestyles."

Others disagreed with such sums. More than two thousand economists, including eight Nobel Prize winners in the field, issued a joint statement in 1997 that argued for "preventive steps" on climate change:

> There are many potential policies to reduce greenhouse-gas emissions for which the total benefits outweigh the total costs. For the United States in particular, sound economic analysis shows that there are policy options that would slow climate change without harming American living standards, and these measures may in fact improve U.S. productivity in the longer run.

Some people, like energy expert Amory Lovins, go even further. He argues that climate protection focused on energy efficiency could even be achieved at a profit: "The confusion originated from imbecilic computer models that simply assumed everything worth buying must have been bought already—i.e., that markets are essentially perfect . . . It doesn't matter how the climate science turns out, because we ought to do the same things anyhow just to save/make money!" As the old joke goes, you could lay all of the world's climate experts end to end and still not reach a conclusion.

How do we make sense of all this? John Weyant, an economist at Stanford University, offers some clues. As head of that university's Energy Modeling Forum, which brings together leading economists from around the world, he gets to peer into the inner workings of many people's economic models. He explains that the wide variance in cost estimates can be attributed not just to ideology but also to differences in seven key assumptions about the future that are built into most models:

> Two of the seven—substitution and innovation—are structural features of the economic models used to make emissions projections. The other five—baseline emissions, policy objectives, policy regime, inclusion of benefits of emission reductions, and discount rate—are external factors or assumptions. Cost projections for a given set of assumptions can vary by a factor of two to four across models because of differences in the models' representation of substitution and innovation processes. However, for an indi-

> vidual model, differences in external assumptions . . . can easily lead to cost estimates that differ by a factor of ten or more.

You don't need to hold a doctorate in economics—or even know what discount rates are—to get Weyant's point: so much uncertainty is embedded in the assumptions fed into economic models that even people with the same ideological bent (never mind spin doctors manipulating the numbers) can come out with costs or benefits that are orders of magnitude apart. There are sure to be some specific climate measures, such as many investments in energy efficiency, that will be unambiguously cost-effective; others, such as stopping the use of fossil fuels immediately, will surely carry a large price tag. However, if you ask the broader question of how much tackling climate change will cost over this century, the honest answer must be that we simply do not know.

As is true with the science of climate change, uncertainty about the economics simply does not justify inaction. Doing nothing also carries costs: the costs of damage resulting from climate change that could have been prevented, for example, as well as the possibility that delaying even small steps now might explode the costs of action later. A more immediate justification for action is abundant evidence that the world uses energy in needlessly inefficient ways. As previous chapters in this book have explained, the centralized electricity distribution model is woefully inadequate: America's aging coal plants convert far less than 40 percent of their fuel into useful energy, and they waste even more energy as the power is shipped to your house; the heat produced at those coal plants is squandered, since it is far from users, so you need to consume more fuel to produce the heat you need at home. As Amory Lovins loves to point out, cars powered by internal combustion engines expend barely 1 percent of the fuel used in order to propel the driver in a forward direction—which, when you think about it, is really the reason you get a car in the first place. Myriad other examples suggest that there is lots of room for improvement.

This matters to the economics of climate change because such grotesque waste of energy is needlessly sending vast quantities of greenhouse gases like carbon dioxide into the atmosphere. But CO_2 has never been thought of as a pollutant, so businesses have never bothered trying to minimize emissions. Shake the trees a bit by giving those businesses a slight incentive to do so, and you might be surprised at how much low-hanging fruit falls to the ground.

The Century's Carbon Challenge

Whatever happens to the Kyoto Protocol, the world's leaders must get cracking on a century-long approach to climate change. So what exactly should such a grand plan look like? First and foremost, it should be global in nature. Climate, after all, is the textbook example of a "commons" problem, in which the transboundary nature of pollution means that the polluter may not end up being the victim of his own pollution. It is also a classic "stock pollutant" problem, meaning that the flow of new pollution matters less than the overall accumulation of pollution in the atmosphere. Since CO_2 lingers in the atmosphere for a century or more, any plan must extend across several generations.

Another good starting point is to remember that climate change per se is nothing new: the climate has been changing continuously through the earth's history. What is new here is that because of the scale and complexity of our industrial societies, we now may have the power to push the climate system beyond a point of no return. While we come to grips with this, it is important to recall the lessons of generations upon generations of our ancestors: mankind must adapt to nature's vicissitudes. This means some obvious things in the rich world, like building bigger dikes and flood defenses. However, since the most vulnerable people are those in poor countries, it also means helping them adapt to the inevitable assaults that will come from rising seas and freakish storms. Infrastructure improvements will help, but increasing the poor world's prosperity (for example, by dropping the shameful subsidies and trade barriers that hinder its exports to the rich world) is probably the best investment.

It's also essential to be quite unambiguous about what the long-term objective of any strategy is. In order to avoid dangerous human interference with the climate system, a growing chorus of scientists argues that we need to keep temperatures from rising much more than 2 to 3 degrees Celsius. Even more important is ensuring that the rate of increase is kept in check. Doing that will require stabilizing atmospheric concentrations of greenhouse gases (whether that target is 550 parts per million of CO_2 or some other level is a second-order matter until the science becomes more precise). James Edmonds of the University of Maryland points out that because of the long life of CO_2, stabilization of CO_2 *concentrations* is not at all the same thing as stabilization of CO_2 *emissions*. That, says Edmonds, points to an unavoidable but still stunning conclusion: "In the very long term, global net CO_2 emissions must eventually peak and gradually decline toward zero, regardless of whether we go for a target of 350 ppm or 1000 ppm." That's right: zero!

The proper objective for climate policy is the eventual transition to a low-carbon energy system. Such a transition might take a century (depending on the CO_2 target), and need not lead to a world without the mobility, light, warmth, and other good things that fossil fuels make possible today. Though breakthroughs in technology (such as fuel-cell cars powered by hydrogen made from solar energy) may well lead the world to carbon-free fuels, we do not have to bet our future on it. The world doesn't have to give up cars or electricity to deal with climate change, for two often overlooked reasons—although both do call for dramatic innovations in the way that energy is generated, transported, and used.

One reason arises from a close look at where precisely the carbon left in the ground is distributed. As Robert Socolow of Princeton University points out, the problem is really coal. Mankind can guzzle all remaining conventional oil and gas resources put together (though not those of unconventional hydrocarbons like

Canada's tar sands) and still meet ambitious climate targets. But if mankind taps the earth's stores of coal, which contain enormous amounts of carbon, the game is up.

Also, it is not emissions of CO_2 but rather *net* emissions of CO_2 that we need to reduce. That leaves scope for the continued use of fossil fuels as the primary source of modern energy if only some magical way can be discovered to capture and dispose of the associated CO_2. Practical and economic solutions are still years away, but scientists are already on the job.

One option is the biological "sequestration" of carbon in forests and agricultural lands. Another promising idea is capturing and storing CO_2 geologically—underground, as a solid, or even at the bottom of the ocean. James Edmonds offers an even niftier idea: "Planting 'energy crops' like switchgrass and using them in conjunction with sequestration techniques could even result in net negative CO_2 emissions, since such plants grow using carbon that was already in the atmosphere." If sequestration is combined with techniques for gasifying coal into hydrogen (see chapter 8 for more details), then that troublesome but abundant fuel could even prove the bridge to a sustainable hydrogen economy.

If a revised Kyoto treaty were robust and encouraged bottom-up innovation, it could unleash such a wave of clean energy technologies. After all, the needlessly dirty and inefficient way we use energy is the single most destructive thing we do to our planet. Whether it is the burning of coal in industrial power plants or the felling of tropical forests, mankind's quest for energy is never-ending. It is also essential to modern life. The key to a sustainable future for mankind is to make that energy use clean and carbon-free as well—and a sensible, long-term climate treaty could be the first step in that direction.

Such grand principles are fine for energy pundits, you might think, but is there any real chance of seeing them applied to the real world in our lifetime? After all, given the bloody Kyoto battlefield, you might think that there is no common ground at all around which to organize the sort of climate strategy outlined above. On the contrary. Step back from the immediate fray and you find that there is broad agreement on precisely the principles outlined above—in a treaty signed by the elder George Bush and, crucially, reaffirmed by his son: the UN Framework Convention on Climate Change (FCCC). This treaty was perhaps the most important outcome of the Rio Summit in 1992, and it remains the basis for the entire international climate-policy regime.

The treaty is global in nature and long-term in perspective. It commits signatories to pursue "the stabilization of greenhouse-gas concentrations in the atmosphere at a level that would prevent dangerous interference with the climate system." Note that the agreement covers greenhouse-gas *concentrations*, and not merely emissions. In effect, this commits even gas-guzzling America to the goal of declining emissions. This broad agreement on first principles is the reason for hope that the world can eventually put the current bitter dispute about Kyoto behind it.

Crucially, the FCCC treaty addresses not only what is agreed but also how it must be achieved: any specific strategy to achieve stabilization of greenhouse-gas concentrations, it insists, "must not be disruptive of the global economy." That is where the Kyoto treaty, which is built directly upon the FCCC agreement, went wrong: its targets and timetables, negotiated on the basis of politics rather than science, proved unrealistic. Any revised Kyoto treaty should rest on three basic pillars.

First of all, governments everywhere (but especially in Europe) need to *agree that emissions reductions must start modestly*. That is because the capital stock involved in the global energy system is vast and long-lived, so a breathless dash to scrap fossil-fuel plants in favor of renewable energy would involve enormous cost. However, it is important for such economic pragmatism to go hand in hand with policy reforms (such as an end to fossil-fuel subsidies) that encourage a switch to low-carbon technologies when existing plants are retired or when new ones are needed.

Second, governments everywhere (but especially in America) need to *send a powerful signal that we are entering a carbon-constrained world.* Whether this is done through carbon taxes, mandated greenhouse-gas emissions restrictions, clever "cap and trade" market mechanisms, and so on is less important than sending a forceful and unambiguous signal. This is the area where Bush's policy waffling has done the greatest harm: American industry, unlike European industry, has no reason whatsoever to think that investments in low-carbon technology will be recognized or rewarded anytime soon. The irony is that even some coal-fired utilities in America are now clamoring for some sort of CO_2 regulation so that they can invest in new plants with confidence.

This points to the third pillar of this low-carbon strategy: *promote science and technology*. That means encouraging basic research and the diffusion of that research into the marketplace. Given that investment in energy and climate research has dropped sharply over the last two decades, governments need to step in directly—and, to his credit, Bush did expand funding for climate science. Rich countries and aid agencies must also research ways to help the poor world adapt to climate change. The best way to speed adoption of technologies, however, is not to pick winners. Far better are policies like carbon taxes that give businesses a strong incentive to embrace climate-friendly technologies.

The only sure way to reduce uncertainty in the climate debate is through sound science. This is especially true if the world makes only gentle cuts in emissions initially. That, observes Tom Wigley, means that by the time we get to mid-century "very large investments would have to have been made—and yet the 'return' on these investments would not be visible. Continued investment is going to require more faith in climate science than currently appears to be the case."

The Past as Prelude

Even a visionary like Churchill might have lost faith in such circumstances. Is there any reason to think that today's petty politicians may yet rise to the occasion? Possibly.

Two decades ago, the world faced a similar environmental dilemma: evidence of a hole in the ozone layer. Then, as now, some early but inconclusive signs emerged that human actions (in that case, the use of chlorofluorocarbons—CFCs—in refrigeration) had unwittingly been contributing to an environmental problem. Then, as now, there was the threat of a disastrous outcome if the problem was ignored. Then, as now, the first impulse of powerful industry interests was to resist forceful action. Then, as now, the chief problem was that only a concerted global response would do: action by the rich world alone could eventually have been undermined by emissions from China and India.

Yet thanks to the leadership of a small handful of countries (led, ironically enough, by America), the Montreal Protocol was signed in 1987. That farsighted global treaty has proved surprisingly successful: not only is the manufacture of CFCs now being phased out, but there are already signs that the ozone layer is on the way to recovery. In September 2002 the United Nations Environment Programme and the World Meteorological Organization released a report that contained good news, with an important caveat: "The world is making steady progress towards the recovery of the ozone layer, with the latest scientific results showing that the total amount of ozone-depleting chemicals in the troposphere (lower atmosphere) continuing to decline, albeit slowly. The findings reinforce the need for strengthened political commitment to ensure the continued compliance with the international treaty known as the Montreal Protocol by developed and developing countries. They also demonstrate the need for greater awareness of the reasons behind this vulnerability, not least a better scientific understanding of the linkages between ozone layer depletion and climate change."

Though the climate problem is infinitely more complex, not least because of the problem of national self-interest identified by Scott Barrett, there are still several lessons to be learned from this success story. First of all, the rich world caused the problem, and it must lead the way in solving it. Second, the poor world must agree to cooperate, but is right to insist on time—as well as money and technology from the rich world—to help it adjust. Third, industry participation is the key: it was only when DuPont and ICI broke rank with the rest of the chemical industry that a deal on CFCs was possible. BP and Shell have similarly broken rank with the world petroleum industry on the climate issue, but the American oil industry remains hostile. Cynics note that DuPont joined the bandwagon only after its researchers had come up with an alternative to CFCs, thus guaranteeing the company's future profits. Perhaps that explains why the oil giants out in front on the climate issue are so busy developing hydrogen technologies.

The final lesson is the most important one. New scientific evidence shows that the threat from ozone depletion turns out to have been much deadlier than was imagined at the time the world decided to act. The Montreal pact was flexible enough to allow negotiators to adjust mankind's response as the science justified; any Son of Kyoto can be designed to be that robust too. Mario Molina of MIT, who shared a Nobel Prize for his pathbreaking research on CFCs, insists that the world must learn from this tale of disaster averted:

> The Montreal Protocol is a model example of environmental cooperation. It established very important precedents relevant to global warming: there are ways for developed and developing countries to work together to successfully address global environmental issues; industry can play a vital role in solving the problem, as long as it perceives a level playing field; and it is essential to provide strong incentives to develop new and cleaner technologies. It offers hope that scientific understanding can once again provide the foundation for decisive action by the international community.

In short, the uncertain nature of a threat like climate change is simply no excuse for inaction. It is time to start a measured, market-led move to an energy system that goes beyond carbon.

Churchill would no doubt have agreed. Though he himself was often enshrouded by the infamous London fog (described in the next chapter, on energy's impact on the local atmosphere), he was still able to see the future remarkably clearly. "The further back you look," he once said, "the further forward you can see."

6

Clearing the Air

SO WHAT COLOR is your sky? That's the question that the California Science Center, an interactive museum in Los Angeles, recently posed to its visitors. The question was inspired by an exhibit by Kim Abeles titled *Sixty Days of Los Angeles Sky Patch (View to the East)*. Here are a few responses from the youngest museumgoers, posted on the museum's website:

> "The sky is blue with white puffs and dark blue at night."
> —*Brittney, age 4*
>
> "The sky is blue because it reflects the attractive bluish color of Krishna (God)." —*Syama S., age 8*
>
> "The sky is yellow. Because it looks dirty, sometimes . . ."
> —*Emily W., age 5*
>
> "The sky is blue and white. The clouds are white the sun makes the sky blue. The sky at night is black because the earth is turning." —*Gabriel R., age 6*
>
> "The sky in California is so gray because of the smog but

if you go to Florida it is totally baby blue and beautiful!"

—Echo T., age 8

"The sky is blue but my mom says it is purple. I think she's right but I can't see it that color. She's a mom."

—Daniel, age 9

"The colors are so inspiring to me that I wish I was a rainbow!!!!!!!!!!!!" *—Alexandra, age 11*

To anyone old enough to remember the air pollution in Los Angeles a few decades ago, such responses must come as a shock and a delight. The problem was the city's infamous smog (a contraction of smoke and fog), which seemed to pounce on it with cat's claws on sunny days—which, in southern California, meant pretty much all the time. Still, some days were worse than others.

"On July 26, 1943, in the midst of World War II, Los Angeles was attacked—not by a foreign enemy, but a domestic one—smog. Smog. SMOG." So begins *The Southland's War on Smog*, an official history of air pollution in L.A. published by California's South Coast Air Quality Management District (AQMD). Accompanying the melodramatic words is a genuinely dramatic picture of a nebulous cityscape from that year described as the first "recorded smog" in the city. The glossy publication goes on to make clear, however, that air pollution was no newcomer to wartime L.A.: "Industrial smoke and fumes were so thick one day in 1903 that residents mistook it for an eclipse of the sun."

The more the local authorities tried to do something about the smog, the more problems they created for ordinary people: restrictions on car use, on wood-burning stoves, and on many other ways that they used energy. The only good thing about the smog, locals used to joke, was that the chemical cocktail made the city's sunsets more spectacular. Yet as those mostly cheerful answers from the youngsters suggest, things are changing. Though city residents will always complain about the pollution, the air is much cleaner than it was back in the 1970s. Old-timers have even started to grouse that sunsets are not as picturesque as they used to be.

Ask people in London about sky color, and you'll be surprised at how many do not know quite how to answer. Many tend to think of the sky in only two shades: rainy, or about to rain. But the question is about pollution, not precipitation. On those rare cloudless days, the London sky is usually an iridescent blue. That is astonishing when you consider that London's air is even more notorious than L.A.'s.

London has had filthy air for more than a millennium. Back in the thirteenth century, the problem was so bad that King Edward I created the world's first pollution-control agency. He also tried to ban coal, the chief culprit behind local pollution, but nobody paid attention. Queen Elizabeth I complained about the air in London, insisting that she was greatly "annoyed with the taste and smoke." She imposed a partial ban on coal burning, also to no effect. John Evelyn, a noted diarist, wrote a century later that the city's "filthy vapour" meant that "catharrs, phthisicks, coughs and consumptions rage more in this one City than in the whole Earth besides." For centuries, the city's coal-fired hearths have conspired with its natural humidity to produce killer London fogs—including one as late as 1952 that lasted a week and led to the premature death of thousands of people.

Though many contemporary Londoners (like their counterparts in Los Angeles) still complain, the air is downright salubrious compared to what Oliver Twist and his real-life counterparts endured. Tony Blair, Britain's prime minister, even declared—on advice from scientific experts—that London's air today is the cleanest it has been since the industrial revolution began. That is hardly the case in the sprawling megacities of the developing world, where air pollution has generally been getting worse. Those inverse trends probably explain why people in rich countries think very differently about air pollution from those in poor countries.

The Same Air That Cleopatra Breathed . . .

Asked what is the biggest threat to the environment, the average European will likely point to global warming. The two shy little boys playing outside Liu Shihua's cigarette shop in Da Shilan, a shabby neighborhood in the heart of Beijing near Tiananmen Square, give a very different answer. Choked by the exhaust from innumerable automobile tailpipes and soot from the neighborhood's coal-fired stoves, one points to the air and exclaims, "It's bad—like a virus!" The youngsters are perceptive: international agencies consistently rank Beijing one of the most polluted big cities in the world, and there are a dozen smaller Chinese cities with pollution much worse. But the boys don't need official statistics to tell them that. Ask them what color their sky is. "Gray!" says one without hesitation. "No, stupid, it's blue," says the other, buffeting his playmate.

Given all the media hype and green hoopla about climate change, it is too easy for people in the rich world to believe that such global scares are the chief environmental problems facing humanity today. They most certainly are not. Partha Dasgupta, a professor at Cambridge University, argues that the current focus on "global, future-oriented" problems has "drawn attention away from the economic misery and ecological degradation endemic in large parts of the world today. Disaster is not something for which the poorest have to wait; it is a frequent occurrence."

He is not exaggerating. Every year in developing countries, at least a million people die from outdoor air pollution. Tragically, perhaps twice that number die from indoor air pollution—usually the result of exposure to stove smoke inside their poorly ventilated homes. This killer strikes women and especially children the hardest, as mothers tend to carry their babies on their backs while hunched over their stoves. Premature deaths and illnesses arising from "environmental" factors (a category that includes both air and water pollution) account for about 20 percent of the entire burden of disease in poor countries. That is bigger than any other preventable factor, including even malnutrition. Yet when was the last time you saw a Live Aid concert or televised charity appeal devoted to helping snuff out cow-dung fires in India?

Poverty experts like Dasgupta insist that tackling local problems must be our top environmental priority: air pollution may not be as sexy a problem as global warming, but it is sure to kill many millions in the coming decades. The problem is so serious that Ian Johnson, the World Bank's vice president for the environment, tells his colleagues (only half jokingly) that he's really the bank's vice president for health: "I say tackling the underlying environmental causes of health problems will do a lot more good than just more hospitals and drugs." Determining how we should do so is a hugely controversial matter that cuts to the heart of the great global debate over "sustainable development."

It is awfully hard to oppose a notion as fuzzy as sustainability. It conjures up heartwarming images of wild animals and untainted wildernesses stretching far into the future. The difficulty comes when you try to reconcile the "development" part of the expression with the "sustainable" part: look a bit closer at that canvas and you will notice that there are no people in it. At the heart of this debate lies the $64,000 question: What do we owe future generations—and how can we reconcile that with what we owe the poorest among us today? Environmental problems like air pollution are central to this debate because the best way to help the poor today—economic growth—can, if handled recklessly, eventually degrade or even destroy the natural environment. That tension explains why ecologists and economists, the two vocations with the most passionate views on sustainable development, have long had diametrically opposed views on the subject.

On the eve of the big Earth Summit in Rio de Janeiro in 1992, the two sides fired salvos that are worth recalling. UNESCO, an agency of the United Nations, offered the following suggestions on what we owe the future: "Every generation should leave water, air and soil resources as pure and unpolluted as when it came on earth

. . . each generation should leave undiminished all the species of animals it found existing on earth." Such a stance invoked the alluring notion of man as but a strand in the web of life, and of the natural order as fixed and supreme. Put Earth first, argued such voices, or risk the future of humanity as well as that of the planet.

Robert Solow, a Nobel Prize–winning economist at MIT, argued at the time that this was "fundamentally the wrong way to go." Our obligation to the future, he wrote, is "not to leave the world as we found it in detail, but rather to leave the option or the capacity to be as well off as we are." That position invoked the seemingly hard-hearted notion economists call "fungibility": that natural resources, be they petroleum or panda bears, are substitutable. By all means save the pandas if it makes you feel better, argued such voices, but do not confuse such action with sustainability—and pay for the luxury yourself.

The ecologists and greens of the Earth First brigade insist that economic growth must be restrained before it does irreparable harm to the planet. The technologists and development economists of the People First camp are equally adamant that tackling poverty through economic growth must remain our top priority. Both insist that theirs is the only true path to the nirvana of sustainable development. As the world's heads of state agreed at the second Earth Summit in South Africa in 2002, figuring out the sustainability puzzle is the most important challenge facing humanity in this new century.

People First or Earth First?

Anyone who has visited Bangkok, Lagos, or Mexico City knows that people in growing economies cannot wait to get onto motorcycles, scooters, and especially into cars. The resultant emissions, when combined with the already nasty soot and sulfur belching forth from coal-fired hearths and boilers, mean that filthy air is now the norm in most of these cities—especially in Asia, which is the fastest-growing region of the world.

Is this cause for alarm? Not really, argue many economists. After all, history shows that poor countries always pollute as they grow richer—just recall how filthy Victorian England was. Such are the growing pains of industrialization. Economic growth not only lifts the masses out of squalor, goes the argument, but in the long run it cleans up nature too. After all, that's what the experience in the developed world shows.

One of the most forceful advocates of the People First view was the late Julian Simon, an academic at the University of Maryland. In *It's Getting Better All the Time*, a sunny book he co-authored with Stephen Moore, Simon pointed to America's undeniable (if largely unheralded) success in combating air pollution as evidence of an inevitable trajectory:

> Economist John Kenneth Galbraith voiced the prevalent attitude in the 1960s and 1970s about economic growth and the environment in his book *The Affluent Society*: "The penultimate Western man, stalled in the ultimate traffic jam and slowly succumbing to carbon monoxide, will not be cheered to hear from the last survivor that the gross national product went up by a record amount." . . . The surprisingly good news is that the economic progress of the last century has not come at the expense of clean air. Rather, economic growth has generally corresponded with improvements in the natural environment.
>
> The national picture on air quality shows improvement for almost every type of pollution—with particularly dramatic declines in carbon monoxide, sulfur, and lead. Lead concentrations have fallen precipitously, by more than 90 percent since 1976 . . . Ambient air pollution levels have been decreasing steadily since the 1970s. Between 1976 and 1997, levels of all six major air pollutants decreased significantly: sulfur dioxide levels decreased by 58 percent, nitrogen dioxides decreased 27 percent, ozone decreased 30 percent, carbon monoxide decreased 61 percent, and lead decreased an overwhelming 97 percent.

Adding credibility to this argument is the fact that the experience in Europe and Japan has been broadly similar to the American success story.

Listen to such pundits for a while and the free-market mantra sounds pretty persuasive. In the long run, prosperity surely is the best guarantor of a better environment. Even so, it is becoming increasingly clear that trade and economic growth will not solve all environmental problems. The British nongovernmental group Oxfam questions this faith in markets by pointing to the unfair conditions under which world trade takes place. It highlights the enormous scale of subsidies lavished by rich countries on industries like agriculture—as well as trade protection offered to manufacturing industries like textiles—as reasons to think the rules of world trade are "unfairly rigged against the poor." That hurts the environment because it forces the world's poorest countries to rely ever more heavily on the extraction and exports of basic commodities—particularly energy-intensive and ungreen activities. The terms under which this trade takes place are so far from ideal, argues Partha Dasgupta of Cambridge University, that they amount to "a massive subsidy of rich-world consumption paid by the world's poorest people."

That is pretty provocative, but now consider this critique of the economic orthodoxy: "Economic growth is not sufficient for turning environmental degradation around. If economic incentives facing producers and consumers do not change with higher incomes, pollution will continue to grow unabated with the growing scale of economic activity." Those words come not from any WTO-bashing, anti-globalization green group, but from the head of research at the World Trade Organization itself.

So growth alone can't solve all environmental problems, but could it be the root cause of those problems in the first place? That is precisely what the Earth First extreme of this debate insists. In fact, it is fashionable among environmentalists in the developed world and grassroots activists in the developing world to argue that trade, globalization, and economic growth are inevitably the enemies of the environment. Thilo Bode, who served for many years as head of Greenpeace International, goes so far as to brand economic growth "simultaneously a false god and a drug":

> From the ecological point of view, dependency on the "growth drug" is based on a serious design fault. Economic growth accompanied by continuously increasing energy and resource consumption does not create prosperity; indeed, it has a destructive effect. In 1950, the gross world product was five trillion dollars; by 1998, it was 29 trillion dollars. Between 1990 and 1998, it increased by six trillion dollars, about as much as between the dawn of civilisation and 1950. Similarly, between 1950 and 2000, wood consumption increased fivefold, grain consumption tripled and four times as much fossil fuel was burned.

The logic behind such arguments is that growth inevitably brings with it urbanization, motorization, and industrialization—all factors that consume lots of energy and so contribute to pollution.

Such voices usually also argue that rich countries—especially the United States, the bête noire of greens everywhere—got rich and stay rich chiefly through the unsustainable guzzling of precious natural resources. This greed, they argue, means mankind will need more than one planet's worth of resources very soon. The WWF (formerly known as the World Wildlife Fund) complains that the "responsibility for this continuing destruction lies with the wasteful over-consumption of rich countries" and that "pressures on the environment from economic growth and demographic change continue to grow."

Listen to such pundits for a while and the deep-green mantra also sounds pretty persuasive. After all, every new factory undeniably brings with it extra pollution. However, economic growth produced by that new factory also unleashes many dynamic forces—ranging from technological innovation to rising consumer welfare—that, over the long run, usually more than offset that extra pollution.

The world's poorest are threatened more by the sorts of hardships we associate with life back in the Stone Age than those of the Internet Age. Experts say the burden of disease related to traditional environmental risks (like indoor air pollution and waterborne diseases) is far greater in poor countries than the problems associated with modern environmental risks (like industrial pollution). The rich world enjoys relatively low levels of both sorts of risks. That suggests that the people in poor countries really suffer from lack of economic development, not from too much of it.

With the arguments so confusing and seemingly contradictory, and with the two camps in the battle so polarized, is there hope for the future? Is there anything we can do for those kids playing on that Beijing street? Happily, a consensus is emerging among economists and ecologists about how to make development sustainable.

Smarth Growth—A New Opiate for the Masses

As the sustainability debate rages on among talking heads, the World Bank offers some advice to the policymakers who actually have to deal with these problems every day: go for "smart growth" that combines economic development with selective forms of environmental protection. This is surprising advice, coming as it does from a pillar of the economic orthodoxy. In fact, a casual reader of its big World Development Report (WDR) 2003—prepared on the theme of sustainable development in time for the Earth Summit II—might think the agency much criticized in the past by anticapitalism groups has itself now turned positively Marxist.

The report suggests that economic growth is simply not enough. It explains that sustainable development is "about enhancing human well-being through time," which means helping people to improve such things as their "sense of self-worth" and not just their net worth. People who think of the World Bank as merely the bankroller of big infrastructure projects that wreck the environment would be astonished: the WDR rejects the view that despoiling nature is an unavoidable and reversible companion to economic growth. Environmental problems are, it insists, "at their root, social problems. The distribution of assets, and of the costs and benefits of different policies, as well as the role of trust, are all critical to the ability of societies to develop competent rules and institutions to address environmental, social and economic problems." The report argues that the "poor and disempowered" must have much greater access to assets if growth is to be sustainable and the world is to avoid social unrest.

Wow. That sounds like a call for the redistribution of assets—perhaps even a call to arms for the workers of the world. Is it? Not at all, insists Zmarak Shalizi, its lead author: "Our arguments are grounded in neoclassical economics and game theory." His team defines a society's assets as including far more than merely land, labor, and capital. Property rights, the rule of law, transparency, and even trust count too. That matters, Shalizi argues, because these assets are already being "appropriated" in many societies by powerful individuals, governments, firms, or bureaucrats. That is why otherwise sensible policies—including ending environmentally damaging subsidies—are often thwarted. "The benefits of such policies are often dispersed, but the costs concentrated, so the policies get captured by a few." Shalizi insists that strong institutions must compensate for this.

The World Bank's arguments are encouraging, for they suggest that economic growth can help both the environment and the world's poorest. It argues that its emphasis on asset allocation need not provoke a bloody proletarian revolution or a repressive backlash from fat cats: future economic growth will create plenty of new assets to spread around to everyone. Over the next five decades, world economic output could quadruple, to $140 trillion, and the world's population rise to nine billion, with humans becoming a chiefly urban species for the first time in history. To many environmentalists, this sounds like a nightmare: think of all those extra factories and automobiles belching filthy emissions into the air! While acknowledging the pressure this will put on the planet, the WDR authors are confident that these trends could herald good news too.

For a start, they note that the real news about population growth is that it is moderating. As a result, the coming years will see a boom in people of working age. Shalizi adds that economic growth deriving from increased production will create an enormous stock of new assets that can—and, he insists, must—be shared with the world's poorest. Even urban migration can be positive, for it means the poor can share in the new jobs, educational opportunities, and housing stock to come. Critics of urban slums and sweatshops all too easily forget the even more miserable existence in subsistence agriculture that drives the indigent to cities in the first place. All this, the World Bank argues, adds up to an opportunity to make growth more inclusive and sustainable. That is a surprisingly hopeful stance, given that sustainability gurus are usually a gloomy bunch.

Why are the men in gray pinstripes now turning green—and maybe even a shade red? A senior World Bank economist explains that unlike in decades past, there is now a lot of good empirical evidence to support the case for smart growth: "Poor countries can abate quite a bit [of pollution] at low economic cost . . . Attention to environmental performance more broadly than just pollution does not systematically lower their economic growth rates." In other words, the Bank is convinced that poor countries can and should try to tackle some environmental problems now, rather than wait till they are rich.

Such thinking is already having a dramatic impact. It was advice from the World Bank that helped persuade China, which for several decades pursued growth at all costs, to start taking its pollution problem seriously. The agency's experts argued to leaders in Beijing that they lost so much potential economic output as a result of pollution (perhaps 3.5 to 7.7 percent of GDP, according to one study that considered health and other costs) that it was actually cheaper to clean up than to ignore the problem. Skeptics say that China's leaders are only paying lip service when they say they will tackle environmental problems, but the evidence suggests that the effort may be sincere. Indeed, from Mexico to the Philippines, many developing countries, without waiting until they are as rich as pristine Switzerland, are now trying to curb the worst excesses of air and water pollution that typically accompany industrialization.

A Killer Haze Hangs over Asia . . .

And not a moment too soon, says Klaus Töpfer. He has watched with alarm as the pollution problems of the poor world have grown in the years that he has headed the United Nations Environment Programme (UNEP). In mid-2002 he unveiled evidence of a particularly worrying new menace that seemed to justify his concern and bolster the case for smart growth. Satellite tracking and computer modeling by UNEP scientists, he announced, had confirmed the presence of a two-mile-thick pollution cloud over much of southern Asia. Töpfer explained the significance of this "Asian Brown Haze":

> The haze is the result of forest fires, the burning of agricultural wastes, dramatic increases in the burning of fossil fuels in vehicles, industries and power stations and emissions from millions of inefficient cookers burning wood, cow dung and other "bio fuels."
>
> More research is needed, but these initial findings clearly indicate that this growing cocktail of soot, particles, aerosols and other pollutants is becoming a major environmental hazard for Asia. There are also global implications—not least because a pollution parcel like this, which stretches three kilometers high, can travel halfway round the globe in a week.

The researchers worried that this seasonal haze was aggravating droughts, altering the winter monsoons, and disrupting agriculture in various ways.

They also suggested that the pollution might lead to "several hundreds of thousands" of premature deaths per year in south Asia alone. More troubling was that it was possible that the phenomenon was even worse over East Asia. Put crudely, Asia could now be suffering from an air pollution problem that could put the infamous London fog and L.A. smog to shame.

Cities like Beijing have thus decided to speed economic history along. Wang Kai, a bright, self-effacing official in Beijing's Environmental Protection Bureau (EPB), welcomed me to his ramshackle office building in early 2002 by observing that it was a good day to talk about air pollution: "We are having our third dust storm of 2002 already in Beijing! Air pollution must be familiar to you, coming from London." He had the unenviable task of organizing the city's drive to clean up air pollution before the 2008 Olympics—unenviable not just because of the scale of the problem but also because China's rulers live in the same city and must breathe that same air. They have staked their reputations on those games being the "Green Olympics," and now they are turning up the heat on Wang and his colleagues at the EPB to deliver results.

As a result, local officials are not waiting for economic efficiency to dictate when businesses and individuals turn to new technologies. Instead, when particular pollution problems are extreme, they are experimenting with the kind of rich-world regulations that mandate cleanups. Dirty factories that do not clean up are being shut down; owners of old cars must pay for the retrofit of pollution-control equipment; local authorities are phasing out cheap coal throughout the central city in favor of cleaner natural gas; and so on. Experts say this dash for clean air will cost China billions of dollars—money that will not be available, of course, for other social needs or to tackle pollution in less politically important Chinese cities. That trade-off is leading some Chinese cities without Beijing's clout or deep pockets to try even more radical experiments.

"We want to turn Taiyuan into a civilized place!" So says Yuan Gaosuo, the gregarious deputy mayor of this grimy industrial city in China's northeast. That seems an odd aspiration at first blush—Taiyuan lies in a part of Shanxi province that is considered to be a cradle of early Chinese civilization. Archaeological sites like the Shuanglin Si monastery and the Jinci Si temple abound. Wutai Shan, one of Buddhism's most sacred sites, attracts millions of visitors each year. Whatever its other deficiencies, surely civilization is one thing that Taiyuan already has going for it? Not according to Yuan: "Without clean air, we cannot consider our city civilized." When experts measured air pollution levels in China's forty-seven biggest cities in 1999 and 2000, Taiyuan was the worst. In fact, with pollution levels at nearly nine times the level deemed safe, Taiyuan was said to have the filthiest air in the world. It was that humiliation that galvanized local officials to tackle the foul soot, smoke, and sulfur spewing forth from the city's many coal-fired industrial plants.

Local environmental officials want to achieve a 50 percent reduction in emissions of sulfur dioxide (SO_2), the chief precursor of acid rain, within five years. More striking than that target (after all, grand proclamations are a specialty of the Chinese leadership) is the fact that they aim to do this using innovative market instruments and not just conventional "command-and-control" approaches to environmental regulation. In April 2002, local and provincial officials agreed to a deal with the Asian Development Bank to develop an emissions-trading system to achieve those SO_2 cuts. That comes on the heels of the success in the southeastern city of Nantong which, with the help of the American green group Environmental Defense, has already completed China's first SO_2 trade. This flurry of activity draws inspiration from America's SO_2 emissions-trading system, also set up in the early 1990s with help from that group, which greatly reduced SO_2 emissions at a cost far less than would have been possible through heavy-handed regulation (see the next chapter for a full discussion of market environmentalism).

The key to trading is flexibility: since firms that find it costly to meet SO_2 targets can buy credits from those that find it cheaper to make cuts in emissions, the overall target is met at the least possible cost. Richard Morgenstern of Resources for the Future (RFF), an American think tank that is advising Taiyuan's government, points out that the city meets two preconditions for trading. First, a small number of large emitters (twenty-six factories, steel mills, and the like) account for half of all SO_2 emissions. Second, those local firms face hugely divergent marginal costs for reducing those emissions—ranging, on RFF's analysis, from $60 to $1,200 per metric ton. That means that there is something real to trade, which in Morgenstern's opinion makes Taiyuan "just ripe for emissions trading." Environmental Defense's top trading expert, Dan Dudek, envisions a nationwide emissions-trading system one day. He is also working on similar schemes in several other countries, including Russia. So great is the interest in developing countries, and so prodigious Dudek's efforts, that one American academic calls him "the Johnny Appleseed of emissions trading."

Hang on a minute, though. All these plans for technological and regulatory "leapfrogging" sound innovative and exciting, but how realistic are they? Even if they have the political will, as the Chinese authorities seem to, can the choking cities of the developing world really hope to defy the economics textbooks and leap ahead to cleaner air before growing richer? The evidence from Los Angeles and London—the pioneers in the world's fight against air pollution—suggests that the answer is an encouraging maybe.

Leaps of Faith

"When I moved to Los Angeles in the 1960s, there was so much soot in the air that it felt like there was a man standing on your chest most of the time." Ron Loveridge, the mayor of Riverside, a city on the east side of the greater Los Angeles basin, knows a thing or two about smog. Because the ocean breezes from the west blow pollution from L.A. over his region, and the mountains that surround the other side of his county trap that pollution, the residents of his "Inland Empire" suffer the worst of the area's air pollution. Loveridge does not despair: "We have come an extraordinary distance in L.A."

That is no idle boast. Four decades ago the city had the worst air quality in America. No one knew why. The man who cracked the puzzle was Arie Haagen-Smit, a Dutch biochemist at the California Institute of Technology who had previously developed perfumes for a living. He noticed that the leaves of agricultural crops near local refineries were discolored or bleached by the local version of air pollution—something that did not happen in other American cities. Similarly, tire manufacturers complained that L.A. smog wore down their tires faster than the dirty air elsewhere. Knowing that the pollution in eastern cities was made up of soot and sulfur chiefly from the burning of coal, Haagen-Smit turned his sights on another culprit: ozone, a form of oxygen that he demonstrated to be dangerous for human health and the environment.

When he explained that this ground-level ozone (as opposed to the ozone found in the stratosphere, which shields humans from the sun's ultraviolet rays) was formed as a result of the reaction of hydrocarbon exhausts and nitrogen oxides with sunlight, big business hit the roof. *The Southland's War on Smog* describes the angry reaction from local industrialists to what was derided by some as the "Haagen-smog" theory:

> Business leaders argued that irritating ozone came not from oil refineries and cars, but from the stratosphere, where ozone descended to the surface of the Pacific Ocean, then was blown over the Los Angeles area by offshore breezes. Haagen-Smit knew that the atmospheric temperature inversion trapping smog close to the Earth's surface would form a barrier preventing ozone from descending from the stratosphere. Finally, in 1954, the Stanford Research Institute, funded by oil companies, showed only low levels of ozone on Catalina Island [just offshore from L.A.], disproving the migrating ozone theory.

Even so, hostility from big business and further scientific mysteries would continue to impede the struggle for clearer skies in L.A. for many years.

Arthur Winer, an atmospheric chemist at the University of California at Los Angeles, explains that tackling smog even after its cause was identified took tremendous perseverance and political will. It turns out that this chemical cocktail behaves in a nonlinear fashion: that is, a 5 percent reduction in the precursors to smog would not necessarily produce a 5 percent reduction in the incidence of smog. Early regulatory efforts met stiff resistance from business interests and faltered as they failed to produce dramatic results.

Clean-air advocates like Ron Loveridge began to despair: "We used to say that we need a 'London fog' here to force change." That tragic episode in 1952, which resulted in at least ten thousand premature deaths, led directly to an act of Parliament that put an end to smoky coal fires (the chief culprit behind London's air pollution) within the city. Though there were noisy objections from some—invoking earlier complaints from George Orwell, who had insisted that a coal fire was "an Englishman's birthright"—the public outcry resulting from the killer fog meant that the reformers in Britain had a stronger hand than the local officials in Los Angeles.

Eventually California officials forged ahead with an ambitious plan that combined a regional regulatory approach with mandates that demanded the use of specific technological fixes aimed at cleaner air. Tackling obvious sources like the burning of trash in backyard incinerators was easy. Turning the screws on stationary sources like refineries and power plants was harder. The nub of the problem, though, turned out to be mobile sources: unsurprisingly, the city's residents did not take kindly to measures designed to curb the use of cars.

Unable to modify behavior much, the authorities had to rely on heavy-handed technological mandates to clear the air. Despite lingering uncertainties, authorities introduced a sequence of pathbreaking but controversial measures: unleaded gasoline, low-sulfur gasoline, onboard diagnostics on cars to minimize emissions, three-way catalytic converters, vapor-recovery attachments for gasoline nozzles, and so on.

The result: the city that issued about 120 "first stage" ozone alerts a year three decades ago did not need to issue a single such alert in the last three years of the twentieth century. Though the population has more than doubled in recent years, and the vehicle-miles driven by car-crazy Angelenos have tripled, ozone levels have fallen by two-thirds. The city's air is undoubtedly cleaner than it was two decades ago.

Lessons from L.A.

Joe Norbeck of the University of California at Riverside praises the state's clean-air campaign: "California, in solving its air quality problem, has solved it for the rest of the United States and the world—but it doesn't get credit for it." Those words are a remarkable vindication of environmental campaigners and policymakers in California, given that Norbeck worked for many years with an automobile industry that fought those environmental laws tooth and nail.

He acknowledges that the problem of air pollution found in the poor world is not identical to the situation in L.A. by any means—poor cities, after all, suffer more from road dust, filthy two-stroke engines (such as those found in auto-rickshaws), and smoky coal fires. Even so, Norbeck is adamant that the poor world's cities can indeed leapfrog ahead by embracing cleaner technologies developed specifically for the California market. He points to China's vehicle fleet as an example: "China's typical car has the emissions of a 1974 Ford Pinto, but the new Buicks sold there use 1990s emissions technology." That matters, because the typical car sold today produces 90 percent less local pollution than a comparable model from the 1970s.

There is one lesson for poor cities like Beijing that are keen to clean up: they can order polluters to meet high emissions standards. And that is precisely what regulators are doing from Beijing to Mexico City. In China, where pollution from sooty coal fires in homes and industrial boilers had been a particular hazard, officials have pushed a switch to natural-gas furnaces. In Mexico City, the "Hoy No Circula" program bans half the cars on the road during pollution alerts based on the last digit of their license plate numbers. The program had its problems: wealthy Mexicans simply bought a cheap, dirty second car. However, officials combated such evasions of the law's intent by exempting new cars and adding incentives for cleaner cars. A study done by Luisa and Mario Molina of MIT suggests that the scheme, which is one part of a broader cleanup effort, is having a beneficial effect by encouraging the turnover of the city's dirty fleet of vehicles. Though levels of ozone and suspended particles are still a problem, levels of other pollutants such as lead, SO_2, carbon monoxide, and nitrogen dioxide are now usually lower than the thresholds defined by international health authorities. And as anyone who has lived in the oxygen-deprived megalopolis will tell you, the air is undoubtedly clearer these days. In fact, visitors sometimes even see the icy peak of the Popocatépetl volcano, which, though close to the city, was usually shrouded from view during the bad old days.

Such technology mandates may be trickier to pull off in impoverished or politically weak cities. City officials must first be willing to endure political fallout. This is hard enough in the power sector, but it is exceptionally difficult in transport. José María Figueres, who served as president of Costa Rica in the 1990s, recalls with anguish the public backlash he endured (even in his famously green country) when he raised prices to cover the costs of removing lead from the country's gasoline: "Their children will benefit, but I was gone from office by the time voters realized it!"

What is more, higher standards for new cars are, in isolation, an incomplete step. Clean technologies like catalytic converters often require cleaner grades of gasoline too. Introducing cleaner fuels, say experts, is an essential lesson from L.A. for poor countries. It will not come for free, of course. And if standards are set unrealistically high, they may well fail: Will Cambodians or Bolivians really be able to maintain cars equipped with the latest onboard diagnostics and other fancy bits of emissions-control technology?

Even so, experts advise developing countries to forge ahead on some specific problems. For example, lead should be phased out quickly. So unambiguous are the health impacts of lead (especially on children), and relatively slight the costs of removing it, that experts agree the trade-off is well worth it. Thanks to a big push from development agencies, and success in a few developing countries, most of the world outside of sub-Saharan Africa is now either lead-free or in the process of getting there.

There is another reason why merely waving a wand and ordering cleaner new cars is inadequate: it does nothing about the vast stock of dirty old ones already on the streets. In most cities, the oldest fifth of the vehicles on the road is likely to produce well over half of the total pollution of all vehicles. That is why policies that encourage a speedier turnover of the fleet make more sense than "zero emissions" mandates affecting only new vehicles.

So there is hope that technology can help the poor leapfrog ahead on at least some environmental problems. Newcomers to urbanization can learn about policies that work—a luxury that London and L.A. did not have in their day. Also, these newcomers can count on aid money and advice from experts at international development banks and think tanks. Most strikingly, if they are willing to pay the price, they can copy the scientific knowledge and new technologies that were developed to clean up pollution in the rich world.

In other words, one important lesson from L.A. is that technological innovation can be the environment's best friend. That is an insight rich in irony, for many environmentalists have traditionally viewed technology as a threat to the environment: machines, went the argument, inevitably mean pollution. Perhaps that is understandable, given that the history of air pollution is so closely intertwined with the history of mankind's use of energy—and that, in turn, is a part of the history of human ingenuity itself.

How Many Planets?

"Nothing endures but change." Heraclitus made that observation more than two thousand years ago, but his wisdom often seems lost on modern environmental thinkers—especially when they think about energy and air pollution. Some invoke scary scenarios that assume that resources—both natural ones, like oil, and man-made ones, like knowledge—are somehow fixed. And others imagine that problems like air pollution will only get ever worse as the world economy roars on and modern technologies proliferate. Yet the history of energy use, especially since industrialization began, suggests that the interplay between man and nature takes place not in a steady state of stagnation, but rather in a dynamic dance of development, scarcity, degradation, innovation, and substitution.

Consider this favorite nightmare scenario offered up by green thinkers these days: What would happen if all of China starts guzzling natural resources as wastefully as America does today? Energy strategists worry about China battling the West for scarce OPEC oil, while environmentalists fret that an energy-hungry China would turn into a global warming nightmare. Or consider the specter invoked by UNEP's Klaus Töpfer: that the nasty haze hanging over Asia today might one day decide to attack America. Such worries echo the sentiments of Mahatma Gandhi, an early advocate of the simple life, five decades ago: "God forbid that India should ever take to industrialism after the manner of the West . . . It took Britain half the resources of this planet to achieve this prosperity. How many planets will a country like India require?"

For anyone concerned about the future of energy and the environment, that question must be confronted head-on. Doing so, however, reveals that there are really two distinct questions wrapped up in that scary scenario: Will the world run out of energy resources? And if not, could the growing affluence of developing nations lead to a global environmental disaster? The key to answering both questions lies in understanding the power of technological innovation.

The first fear is the easier to refute. Early classical economists such as Malthus, Ricardo, and Mill all worried that scarcity of natural resources would lead to diminishing returns on economic effort, and eventually an end to growth altogether. Early conservationists a century ago also embraced the notion of scarcity as justification for government intervention in the management of resources. A few decades ago, the Limits to Growth camp of ecologists and intellectuals reinvigorated this line of thinking, inspiring fears that oil and food might soon run out.

And yet there are now more proven reserves of petroleum than three decades ago; more food is produced on earth than ever before; and the last decade has seen the greatest economic expansion in history. How was this miracle possible? The answer lies in understanding that the stock of available resources is not constant. Fears of oil scarcity led to investment that discovered better and cheaper ways to produce more oil; it also led to inventions like more efficient engines that burned less oil, and to the substitution of some petroleum-based chemicals with newly developed alternatives. Ultimately, it may even lead us to abandon oil altogether as the hydrogen-fuel-cell economy gathers speed.

A look back at the history of fossil fuels confirms the central role of technological innovation. The modern economy was built first on coal and more recently on oil. But these fuels didn't appear out of a hole in the ground. They really sprang forth as a result of human ingenuity.

Coal, Oil, and a Dash of Inspiration

For most of history, mankind's chief energy source has been wood, and plenty of it. During the Middle Ages, Europe was densely populated with trees but not people. As a result, fuel was cheap. But thanks to a booming economy and extensive deforestation, by 1700, England was facing a major energy crisis. Or at least, much like today's America, it thought it was. Pundits rang the alarm bells about the soaring cost of wood for heating and for the iron industry; the price of charcoal doubled in real terms between 1630 and 1700.

Yet to the surprise of many, the world did not come to an end. What actually happened was that growing demand produced supplies of energy from unexpected quarters. Enticed by rising prices, entrepreneurs rose to the occasion. They found a way to bring to market a substance that had largely been overlooked until then: coal. That was a turning point in history, for without coal there would have been no industrial revolution. Waterwheels and windmills had been used for ages, but they could not possibly have produced enough energy for the revolution that was about to come.

By 1850 or so, however, the country's leaders began to fear another energy crisis. This time they thought they might run out of coal. A leading economist of the age, W. S. Jevons, crunched the numbers and was convinced that Britain would run out of coal by 1900. And analyzing what little information he had at hand, he added that petroleum simply could not compensate. Wrong again. The perceived shortage of coal coincided with a global shortage of whale oil, which for centuries had been the lighting fuel of choice. It was the confluence of these forces that inspired the race among entrepreneurs in America and Europe to seek out alternatives.

Guess what they found? That is not to suggest that using the gooey stuff had never been heard of before. Quite the contrary. As far back as 3000 B.C., asphalt was used to caulk Sumerian boats and to waterproof baths for the rulers of the Indus Valley. By A.D. 100 or so, the Romans were burning crude oil as a fumigant against caterpillars. By the tenth century, Arabs had figured out how to distill it; the streets of Cairo were lit by oil. Marco Polo noted in 1272 as he passed by Baku, on the Caspian Sea, that oil lamps were common. At the end of the eighteenth century, Britain's first envoy in Borneo found a flourishing oil industry: some five hundred hand-dug wells supplied light and heat for Bornean homes.

A reasonable question is why petroleum didn't emerge earlier as a serious energy source. The answer lies, yet again, in the interplay of population and technology. Oil oozing out of rocks or dug out by hand was usually enough to meet the modest needs of those who lived nearby. But Europe, where the early arrival of industrialization sent energy demand soaring, didn't have much oil. So proto-pundits assumed it wouldn't play a big role in the new economy. Yet again, they ignored the power of human ingenuity.

The Prize, Daniel Yergin's Pulitzer Prize–winning history of oil, describes the forces that gave birth to the modern oil industry:

> The hopes pinned on the still mysterious properties of oil arose from pure necessity. Burgeoning populations and the spreading economic development of the industrial revolution had increased the demand for artificial illumination beyond the simple wick dipped into some animal grease or vegetable fat, which was the best that most could afford over the ages, if they could afford anything at all. For those who had money, oil from the sperm whale had for hundreds of years set the standard for high-quality illumination; but even as demand was growing, the whale schools of the Atlantic had been decimated . . . Cheaper lighting fluids had been developed. Alas, all of them were inferior. The most popular was camphene, a derivative of turpentine, which produced good light but had the unfortunate drawback of being highly flammable, compounded by an even more unattractive tendency to explode in people's houses . . . There was that second need as well—lubrication. The advances in mechanical production had led to such machines as power looms and the steam printing press, which created too much friction for such common lubricants as lard.

All this added up to crisis in the minds of some, but to others it smelled like opportunity.

Inventors and financiers on both sides of the Atlantic set out to find the magic substance that would grease or light the world—not initially, it must be noted, to power it—and make them rich. This blunderbuss of innovation ultimately led them to the fuel of the twentieth century. And not just a fuel but, indirectly, a technology: the internal combustion engine. That combination of petroleum and automobile has to be one of the most economically important and personally liberating innovations of the last millennium—as well as one of the biggest sources of air pollution.

In *Scarcity and Growth: The Economics of Natural Resource Availability*, two economists affiliated with Resources for the Future sum up the astonishing role that innovation has played in the history of natural resources: "Recognition of the possibility of technological progress clearly cuts the ground from under the concept of Malthusian scarcity . . . decades ago Vermont granite was only building and tombstone material; now it is a potential fuel, each ton of which has a usable energy content (uranium) equal to 150 tons of coal. The notion of an absolute limit to natural resource availability is untenable when the definition of resources changes drastically and unpredictably over time." The irony is that those words were penned by Harold Barnett and Chandler Morse in 1963, long before the Limits to Growth bandwagon got it wrong.

Of course, that does not answer Gandhi's question about how many planets humanity will need, which is really a way of worrying about the intensity of resource use—especially energy use. If China's economy were to transform overnight into a clone of America's, then it could indeed lead to an ecological nightmare. Extrapolating America's per capita emissions of carbon dioxide, a billion eager new consumers would almost certainly wreck any attempts at climate stabilization. What is more, if America's broader pattern of resource use were replicated lock, stock, and barrel across the developing world today, many small local environmental crises could well stress local ecosystems to the point that the quality of human life would become unbearable. However, thanks again to technology, there is good reason to think that things will not unfold quite as doomsters fear.

"The rest of the world will not live like America!" insists Jesse

Ausubel of New York's Rockefeller University. His point is not that poor people around the world do not covet the basic necessities and creature comforts that Americans have. Of course, they do. His point is that the economic growth that will improve their lot will not come overnight; rather, it will take decades.

That is a crucial distinction, because that extra time means developing countries have the chance to embrace new technologies that are more efficient and less environmentally damaging than the ones in use today. In fact, Ausubel goes further: he argues that when they get to the level of economic might per person that America enjoys now, their ecological footprint will be far lighter. He points to the silent but powerful history of "dematerialization" and "decarbonization" as evidence. Viewed across very long spans of time, it becomes clear that economies—whether America's or China's—use ever fewer material inputs per unit of economic output.

Historically, the driving force behind this was not environmentalism, but efficiency. Short-term scarcities of particular resources typically drive technological innovation that leads to ever more efficient substitutes. Clunky metals have been replaced over time by lighter, more versatile substitutes, such as aluminum and especially plastic. That is why, Ausubel concludes, "When China has today's American mobility, it will not have today's American cars." Rather, he insists, Chinese consumers will enjoy the cleaner and more efficient cars of tomorrow. Consider Joe Norbeck's observation about how clean even today's Chinese cars are compared to the models of a few years ago, and it seems the developing world might yet tame the problem of automobile emissions after all.

So Nothing to Worry About, Then?

History provides a profound and reassuring lesson for us. That still does not mean technology leapfrogging is a magic bullet. Kenneth Arrow, a Nobel Prize–winning economist at Stanford University, notes that being technologically "backward" allowed Japan to invest in the newest capital stock after the Second World War and thereby gain a competitive advantage over richer rivals. However, Arrow worries that most poor countries today lack the intellectual capital, technological prowess, and physical infrastructure to repeat that trick. Luisa and Mario Molina of MIT, who have studied the matter closely, add that technology is less important than the institutional capacity, legal safeguards, and financial resources to back it up: "The most important underlying factor is political will." Installing fancy pollution-control equipment means nothing if it is not accompanied by a culture committed to making it work as well.

Ruth Greenspan Bell of Resources for the Future also argues that there are limits to leapfrogging—especially when those leaps venture beyond mere technologies to entire regulatory structures. She is skeptical about efforts to introduce innovations like emissions trading to developing countries:

> Pollution is a by-product of life. And it is produced continuously in a lot of different ways and places. The effort to control pollution requires a lot of people to do a lot of things that are costly and inconvenient. It is unrealistic to think that each of these will learn at the same rate. Or even that they take the same lessons from similar experiences. Some industries in the U.S. supported market-based instruments and some opposed them.
>
> To me, it is more realistic to think of environmental protection as a herding process. Get everyone moving more or less the same direction. Use vigorous enforcement to turn around the ones who try to run backwards . . . It's wrong to make the people of the developing world into guinea pigs for theories that haven't even been fully worked out in the developed world.

If you really watch frogs leap, she observes, "they don't just leap forward . . . they leap sideways, backwards and every which way."

And overambitious attempts at leapfrogging, she cautions, could prove precisely the same.

Defenders of market-based approaches in poor countries accept such caveats but retort that they do not add up to a case for giving up altogether. One expert neck-deep in such reforms puts it this way:

> Such objections raise general questions about the ability of developing countries to develop the required institutional infrastructure for market-based instruments like trading. Of course, developing countries also have difficulty establishing the needed institutional infrastructure for command-and-control policies too. In fact, it's not really clear the institutional burdens are that much greater for market-based instruments than for command-and-control. The real problem, in my opinion, is that developing countries often lack the political will and the institutions to impose strict standards of any sort.

That points to the most important conclusion for pollution fighters everywhere: innovation is a powerful force, but the broader institutional and legal context—and especially government policy—matters too.

Policymakers can draw two important lessons from this. First, don't do stupid things that inhibit innovation. Second, do sensible things that reward innovators for developing technologies that enhance, rather than degrade, the environment. Though long-term global trends seem to suggest some sort of invisible hand pushing innovation forward, it is in fact the result of societal choices. As even a techno-optimist like Ausubel concedes, "There is nothing automatic about technological innovation; in fact, at the micro level, it's bloody." He notes that there are plenty of examples, from medieval China to twentieth-century Argentina (and perhaps to tomorrow's Europe, given the continent's hysterical and unscientific rejection of genetically modified foods), of societies that turned their backs on technology—and stagnated as a result.

Though free-marketeers will howl in protest, there is one other reason why government matters: technological change does not always go hand in hand with environmental improvement. It often does, as the long history of dematerialization shows, by sparing resources. And sometimes it does not. Consider the development of sport-utility vehicles for a moment: very innovative, but hardly green. Now imagine that those same SUVs come out next year equipped with engines that spew forth absolutely no nasty emissions of any kind. Surely that's the sort of radical solution that will help Wang Kai and his colleagues crack Beijing's notorious air pollution, right?

The Trouble with Zero

I wasn't planning to rent a battery-powered car when I arrived at the Los Angeles airport. Yet there I was in the rental line at LAX, being taunted by the slogans "Electric Vehicles are now here!" "EV rentals are affordable" "Save time and money!" My mental picture of such enviro-mobiles was closer to a golf cart than anything I'd ever seriously consider driving on the L.A. freeways. Yet the "zero emissions" vehicles they were offering at this joint were pretty snazzy: a sporty model that could do 0 to 60 miles per hour faster than a Corvette, and a beautiful Toyota RAV sport-utility vehicle. Save the planet in an SUV—why not, I thought?

As the purpose of my visit was, in fact, to research the smog story presented in this chapter, I decided I'd better check one out. I remained skeptical. After all, where do you recharge these darned things? The agency's friendly executive responsible for zero-emission vehicle rentals explained to me that I had nothing to worry about. Unlike the first generation of battery cars, her vehicles recharge pretty quickly and have quite a long range. Plus, she gave me a hefty guidebook showing that the L.A. region is chock-full of recharging stations where you can get charged up for free. "In fact," she confided, "you can also use the high-occupancy vehicle lanes by yourself, even without a car pool. And you get free parking right up front at a lot of garages and public places. That's the main reason people rent these cars—but if it helps the environment, I say it's still OK!"

She deftly took me through my full itinerary, from my lodgings in Santa Monica all the way out to the AQMD headquarters in Diamond Bar and Mayor Loveridge's office in Riverside, and she showed how many convenient recharging stations there were on my route. As it turned out, because I was visiting people actually involved with the fight for clean air, all of them had recharging stations right in front of their buildings. I couldn't possibly go wrong, I thought to myself. I rented the SUV and took it for a spin up into the Hollywood Hills to catch the sunset over Malibu. Despite what the old-timers say, it's still pretty breathtaking. I was particularly pleased that my green machine zipped up that hill while releasing no nasty emissions whatsoever. It all seemed too good to be true.

Well, it was. The vehicle proved to have a much shorter range than I thought it would—closer to 50 miles than to 100. The fact that I sped along at 80 mph in those empty HOV lanes might have drained the battery faster, but only certain highways had that lane; more often, I was crawling along in traffic like everyone else. And most of the time, I was going nowhere at all, since my vehicle kept running out of power. Charging proved the biggest nightmare. There were plenty of chargers around, but some were of the wrong sort; others were locked or nonfunctional. And rather than the "pretty quick" recharge, my useless battery took more than five hours for a full charge. As a result, my entire visit turned into a fiasco of delayed or missed appointments, apologetic cell-phone calls, and panicky exits from the highway to obscure malls and commuter-rail stations in search of a charger.

It was well into the night on my last day when I made it to Arthur Winer's office at UCLA, hours late for my appointment. I was riding on fumes, as it were, as I was just about to run out of charge for the second time that day. He was unbelievably gracious: "Don't worry, it happens all the time. I once even lost a VIP from

Washington who was due to speak at a conference because she ran out of charge too!"

Before we could discuss any serious atmospheric chemistry, I had to ask him one simple question: How in the world did L.A. end up with these ridiculous clunkers? He explained to me that the problem was the overzealous application of the same policy that had produced dramatic results in the fight against smog: technology mandates. In this case, the culprit was the state's Zero Emission Vehicle (ZEV) mandate. Well-meaning bureaucrats had been convinced some years ago that battery cars would be the technology of the future. In order to push this cure-all to their pollution problem, they issued the order demanding that manufacturers shift over time to cars emitting absolutely no emissions at all. That, they were convinced, would boost battery cars into commercial viability.

Automakers grudgingly invested billions of dollars, but the technology never took off. As chapter 8 explains, batteries—unlike fuel cells, a different technology for powering zero-emission cars that the pollution regulators in California had not initially anticipated—have inherent limits imposed by the laws of physics and chemistry. Even massive tax incentives and manufacturers' subsidies didn't make much difference. The Toyota RAV I rented, for example, sold for over $40,000—more than double the price of a conventional gasoline RAV. Though consumers could claim $13,000 in state and federal tax credits for buying that model, there were still very few takers. Such dismal economics explain why Ford announced late in 2002 that it would shut down its TH!NK electric car venture; General Motors decided to give away thousands of its electric cars, since consumers—even ones who care about the environment—simply refused to buy them.

As that fiasco on the highways and byways of L.A. makes abundantly clear, governments need to tread lightly when it comes to mandating green technologies. Smart growth is clearly a promising new paradigm for sustainable development, and government intervention does make sense in tackling such particular environmental problems as smog. And as Arthur Winer points out, the same ZEV mandate that produced those dead-end battery cars did in the end promote the development of gasoline-electric hybrid vehicles like the Toyota Prius and Honda Insight. These new cars, which are much cleaner than conventional models, are commercial successes.

That suggests it's wise for governments always to consider a range of policy options when trying to nudge innovators in greener directions. The underlying principle remains the same, however: to send a powerful signal to the market that the environment matters. And as the next chapter explains, there is no more powerful signal than price.

7

Adam Smith Meets Rachel Carson

"WE WOULD LIKE to invite you to join us for dinner and a discussion about the links between national energy security, sound energy policies and a healthy economy and environment." So began the simple letter of invitation to the dinner party I found myself at one cold winter evening. The venue was an elegant apartment on Manhattan's Upper East Side, overlooking Central Park, and the evening's chosen topic was close to my heart. The hosts, Paul and Joanne, were strangers to me, but they proved to be very warm and welcoming. They were also quite generous with the aperitifs. Perhaps as a result, the other guests, some of them also journalists, were in good spirits as we got to know each other before the meal. The hosts were quite witty too: they like to inform dinner guests ahead of time that "it should be a lively evening—guns must be left in the lobby." So why was I so nervous?

The hosts were none other than Paul Newman and Joanne Woodward. Of course, their celebrity had little to do with my desire to attend: the event carried quite a lot of journalistic importance. The topic of discussion—whether Congress should raise automobile fuel-efficiency standards—was one of the hottest political controversies of the day. What is more, Newman is a lot more interesting and thoughtful than the garden-variety Hollywood tree-hugger: he is famous for racing sports cars, after all, and he has even openly supported nuclear power. In short, the evening had all the makings of a good story.

All that is true, but I'll come clean: Paul Newman has been an idol of mine for ages. I was nowhere near New York that week. I even took a cab from Kennedy Airport to Newman's apartment directly and staked out the joint for a while rather than risk turning up late because of traffic. I had chatted with a real superstar once before. Well, sort of. Back in 1994, as I was standing in front of the Biltmore Hotel in Miami, Cindy Crawford stepped out of a limo, walked right up to me, and said those three magic words every man longs to hear: "You the bellboy?" As I could manage only a few unintelligible croaks, she slipped away into the night.

Resolved not to let my encounter with Paul Newman go the same way, I racked my brain on the transcontinental flight to New York that day for brilliant and provocative one-liners. Of course, when the hosts turned to me that evening, none of those witty lines came to mind. Out of desperation I told them instead about the experience I had just had in L.A. with that cursed battery-powered rental car that kept running out of juice. Before Newman or Woodward could respond, a man standing beside me started chuckling: "Ha! That's why governments should set the environmental bar and let the market pick the winners."

At first I thought that these fighting words had come from the free-market evangelist from *Forbes* magazine I had just been chatting with. In fact, they were uttered by Fred Krupp, the head of Environmental Defense. I turned to him, and he went on: "There's a problem when markets aren't engaged. The Zero Emissions Vehicle requirement forced new technology, which is good, but it couldn't force it to be widely used or spur developments of competing ideas." On its face, this statement was astonishing: the head of one of America's leading green NGOs questioning aggressive government environmental mandates and praising market forces.

What's going on? After all, the conventional view among greens is that capitalism is the enemy of the environment. Philip Shabecoff, a respected former *New York Times* reporter, summed up this view in his book *Earth Rising: American Environmentalism in the 21st Century*:

> Market economics does not deal well with issues of ethics or morality and, in fact, is largely blind to the broader world that encompasses human existence. As Frances Moore Lappé pointed out, the market economy responds to purchasing power, not human needs . . .
>
> Look at the poisoned brownfields across America, the naked hills and sterile streams of Appalachia, the blasted, abandoned blocks of tenements in our inner cities; the barren strip malls of suburbia, the productive farmland disappearing under ticky-tacky housing developments, fields made hard and barren by excessive use of chemical fertilizers and pesticides. Private ownership may be a motivation for safeguarding property in some cases—but it depends on the kind of property, who owns it, and what the owner wants from it. And the free market fails miserably and absolutely to protect our common property—the air, the atmosphere, the health and safety of city streets and tenements, the lakes and rivers and oceans, and the forests and watersheds, the abundance and diversity of life.

So is Krupp a sellout or a con man? Actually, he and his band of eco-radicals in pinstripes may well be applauded as green pioneers one day. That is because they spotted very early on the wisdom of harnessing the power of markets to do good for the planet. Environmental Defense's pathbreaking efforts helped launch America's sulfur-dioxide trading system (described later this chapter), which is widely credited with tackling the acid rain problem at much lower cost than originally forecast. More recently, it has been pushing for such an innovative, market-based approach in tackling climate change too. Yet for most of the last two decades Krupp's group and other advocates of market-based environmentalism have been denounced as traitors by their fellow greens.

Roll Out the Pork Barrel

Big businesses are undoubtedly big polluters. Pollution is an inevitable by-product of economic activity, and it does not appear on any company's balance sheet of profit calculations. Even so, is it really right to link free markets so casually with fat cats and a foul earth? The truth is, big business rarely supports genuinely free markets. The entire history of corporate America—as in Europe, Japan, or the developing world—is really one of corporatism and cronyism. Industries of every stripe talk publicly about favoring free trade and competitive markets, but behind closed doors they lobby their allies in government intensely for subsidies, tariff protection, state aid, and all manner of anticompetitive intervention in the marketplace. That allows lazy, uncompetitive, technologically backward firms to produce shoddy goods and—surprise, surprise—needlessly high amounts of pollution. Billions of dollars are wasted in this way on well-established American industries like sugar, steel, textiles, and agribusiness. Genuinely free markets would still produce pollution, of course, but they would also unleash powerful competition that would check the market power of dirty and inefficient firms in these and other industries. That is the last thing crony capitalists ever really want—and why free markets ought to be considered the greatest ally of environmentalists in their struggle to rein in corporate polluters.

If that argument sounds far-fetched, consider this question from a green with impeccable credentials: "Why are there so few price signals in America's environmental laws?" asks Carl Pope, the boss of the Sierra Club, angrily. "Businesses are simply not interested in paying the true cost of pollution!" He points to what he considers a sorry litany of handouts, subsidies, and corporate welfare: no tax on carbon emissions; no tax on gas-guzzlers; giveaway prices for mineral rights, grazing, water, and timber drawn from federal lands. He reckons that these handouts are not at all due to market forces, but rather to the peculiar politics of pork.

Fred Krupp agrees wholeheartedly. It might seem that businesses, which currently bear the brunt of costs imposed by today's "command-and-control" system of environmental regulation, would be leading the movement for market reforms. The reason that is often not the case, he explains, is that "vested interests in every industry want to defend their existing position or to promote particular green technologies that they manufacture." Such firms (say, a manufacturer of a particular sort of scrubber) will fight tooth and nail to preserve federal mandates that demand the use of that particular technology rather than promote flexible approaches that would allow the market to choose the best, cheapest approach to solving the problem.

A study conducted by the World Economic Forum and led by Daniel Esty of Yale University underscored that crony capitalism—as opposed to the sort based on competitive markets—is bad for the environment. Esty and his colleagues have sorted through 68 separate variables that they reckon influence environmental sustainability (ranging from corruption to aquifer depletion to sulfur dioxide in the air); they then devised 20 core indicators for an Environmental Sustainability Index (ESI) comprised of those variables, which they weighted equally for the purposes of the country rankings.

At the end of all the number crunching, they found two factors that more than any others influenced the environmental sustainability of countries. The most powerful is income per head: unsurprisingly, Norway fared better than Haiti in their rankings. However, the second big factor was corruption: the less corrupt a country was, whatever its income level, the more likely it was to score high in the rankings. This surprising evidence helps explain why cronyistic Belgium, one of the richer countries in Europe, scored lower than a number of developing countries on the ESI ranking. Why? Esty reckons that corruption (as measured by Transparency International, a well-known anticorruption group) is "a proxy for lots of other things, such as the rule of law and the protection of property rights, that have an important influence on how individuals treat natural resources." In other words, crony capitalism is not just bad for the poor—it is bad for the environment too.

The good news is that environmentalists—who themselves have been guilty of lobbying for white elephant projects as long as they were in favored areas like wind or solar power—are now beginning to see the wisdom of pushing for openness, transparency, and genuinely free markets instead. The Sierra Club's Carl Pope even went so far as to join hands with Ed Crane, head of the libertarian Cato Institute, to fight the pork-laden energy bill that the U.S. Congress contemplated in mid-2002. The unlikely allies penned this truly astonishing opinion piece in *The Washington Post*:

> As members of the House and Senate wrangle over how to hitch together their versions of an energy plan, the two of us—one a committed conservationist, the other an advocate of free-market principles—find ourselves in a rare moment of harmony . . . The legislation does nothing to improve the efficiency of energy markets or to remedy any market failures. In fact, it makes matters worse by further distorting these markets with billions of dollars of taxpayer subsidies and other handouts to well-connected energy industries.
>
> Conservative legislators have conveniently forgotten the economists' admonition that if a technology is economically competitive, no public subsidies are necessary, and if a technology is not economically competitive, no amount of public subsidy or special favors will make it so. Liberal legislators, on the other hand, are so hypnotized by the

> handouts to their favorite industries that they overlook the far greater sacks of largess bestowed on their economic competitors.
>
> We would hope it's not too late for something better. Devotees of Adam Smith and Rachel Carson should join together to propose an alternative bill, one that would simply strip away all energy subsidies and preferences from the budget and the federal tax code.
>
> Environmentalists would be happy if renewable energy sources and energy-efficient technologies were just allowed to compete with the fossil fuels industry on a level playing field. The only way to level it is to end the ever-escalating arms race of corporate subsidies that guarantee green technologies will never win, no matter which party is in power. Likewise, many economic conservatives are more interested in freeing energy markets than in rigging them, but they've lacked the ability to fight the pork-barrel crowd in Washington.

What an evocative image: Adam Smith, the hero of free-market economics, and Rachel Carson, the crusading environmental writer who inspired a generation of passionate greens, coming together to save the earth sensibly. This is nothing short of a manifesto for an entirely new approach to nature: market-based environmentalism.

Will it really happen? There are promising signs. Consider that dinner at Paul Newman's place, where Fred Krupp of Environmental Defense surprised me with his pro-market stance. The conversation at my table started off as a discussion of America's Corporate Average Fuel Economy (CAFE) law, a heavy-handed federal statute that dictates fuel-efficiency standards to carmakers, but it ended up as a debate over a radically different, market-based proposal: carbon taxes. While some questioned the political viability of new taxes, environmental experts agreed that such taxes would be a less market-distorting and more efficient approach to tackling fuel efficiency. The next green revolution is already getting under way.

Pinchot to Pinochet: People, Property Rights, and Pollution

"The foresighted utilization, preservation, and/or renewal of forests, waters, lands, and minerals, for the greatest good of the greatest number for the longest time." That, insisted Gifford Pinchot, should be the proper goal of all environmentalists. Pinchot, who headed America's forest service nearly a century ago, was a pioneer whose approach to conservation continues to inspire countless Americans.

Sadly, that is chiefly because the approach to environmental protection embraced by the world—and especially America—has often rejected Pinchot's call for wise use and management of resources. Instead, governments everywhere have embraced unwieldy centralized approaches involving large bureaucracies and heavy-handed technological mandates. In America, these came in the shape of a wave of federal environmental laws passed three decades ago, around the time of the first Earth Day. The Clean Air Act and the Clean Water Act are the most notable of them. Much of the rest of the world followed the same path in subsequent years.

This top-down approach to environmental protection has long been a point of pride for groups like the Natural Resources Defense Council (NRDC), one of America's most influential environmental outfits. And with some reason, for it has had its successes: air and water in the developed world are undoubtedly cleaner than they were three decades ago. This has convinced such groups to defend the green status quo stoutly. Jerry Taylor of the Cato Institute exclaims in despair, "The NRDC is the Vatican of command-and-control thinking!"

Bearing that in mind, consider these words uttered recently by a leading environmentalist: "Thirty years ago, the economists at Resources for the Future were pushing the idea of pollution taxes. We lawyers at NRDC thought they were nuts, and feared that they'd derail command-and-control measures like the Clean Air Act, so we opposed them. Looking back, I'd have to say this was the single biggest failure in environmental management—not getting the prices right." Now consider who said these words: Gus Speth, head of Yale University's school of forestry and environmental studies (which, as it happens, was founded by Gifford Pinchot). Speth was one of the founders of NRDC, and he later went on to lead other high churches of the environmental establishment including the World Resources Institute and the United Nations Development Programme. His mea culpa is akin to the Pope questioning the Bible.

People like Speth are conceding that the legacy of the "mandate, regulate, and litigate" approach is decidedly mixed. It has undoubtedly brought environmental gains. Yet even as the air was getting cleaner, the metaphorical atmosphere was being poisoned by the confrontational strategy. For decades the prevailing attitude of green groups (like NRDC) and of governments (especially America's) has been one of hostility toward industry, and the resultant policies have encouraged endless lawsuits even as they stifled innovation and cost-effective solutions. This old-fashioned approach is also not flexible enough for the more complex challenges of the future.

Daniel Esty, Speth's colleague at Yale, also thinks it is high time for a change. He explains that the earlier generation of laws "often looked disapprovingly at human activities and economic growth because of their harmful pollution side effects, which were thought inescapable." That disapproving attitude is still prevalent, as any of the protesters at the next big meeting of the World Trade Organization or the World Bank will tell you. What is more, enforcement efforts have generally failed to distinguish firms with good records from those with bad records. Unlike the tax authorities in most developed countries, which focus their scarce resources on a few suspicious cases each year, America's EPA typically spends much of its compliance money going after firms that are generally good citizens. "Prospects for further progress on the same path are limited," insists Esty. He has been involved in developing next-generation environmental reforms that encourage incentive-based, decentralized, and market-oriented policies.

This initiative has attracted much attention from top policymakers—including Christine Todd Whitman, George Bush's first administrator of the EPA—and has sparked off a flurry of state-led experiments. When she was governor of New Jersey, Whitman encouraged innovative schemes, including the speedy rehabilitation of "brownfield" toxic waste sites that remained in a dangerous, unproductive limbo because of legal and bureaucratic wrangles over how to clean them up.

Wisconsin is another state that is pioneering reforms that distinguish good corporate citizens from those with a bad record of compliance. In 1997 it enacted a five-year experimental law that allowed state officials to sign voluntary, enforceable agreements with several firms that would deliver environmental results that were beyond the legal minimum. In 1999, following difficult negotiations with the EPA, Wisconsin became the first state to sign a regulatory innovation agreement with the agency—a necessity to reduce business anxiety about federal second-guessing that plagued early sign-ups.

Two early examples from Wisconsin suggest that flexible approaches really can do better for the planet. In the first instance, a cardboard-packaging firm used the law's flexibility to install equipment that captured methanol in a liquid rather than gaseous state. The new technology resulted in the capture of 1.3 million pounds of methanol emissions in a year, compared to only 160,000 pounds a year under command-and-control rules that prohibited the technology. Second, in less than a year, a fossil-fuel energy-generating facility was allowed to extract 22,500 tons of previously landfilled fly ash, re-burn it for recoverable energy, and sell it as an ingredient for cement. That reduced CO_2 and other emissions from cement production and saved the energy equivalent of fifty railcars of coal. Before the reforms, rigid rules governing solid waste had prohibited the recovery. Such successes have encouraged other states to follow Wisconsin's lead.

The Green Dilemma

Yesterday's failed ambitions, today's hefty price tag, and tomorrow's even trickier targets are driving this new green thinking. For a start, the command-and-control system has not solved all of yesterday's problems. While air and water quality has improved, many other environmental problems—ranging from waste management to hazardous releases to fisheries depletion—have not. Crude tactics like regulatory mandates are better at catching the big polluters than nabbing the many smaller businesses, farms, and individuals responsible for much of today's pollution.

Second, federal regulations are usually inefficient and untargeted, so even the gains so far have come at a needlessly high price. Federal mandates specify technological fixes in order to get to preordained outcomes, with little consideration for local environmental conditions or the "marginal cost" of pollution abatement at individual firms. It may come as a surprise that the most commonsense tool of economics, cost-benefit analysis, has often not been utilized. Indeed, in some cases, America's Congress has expressly forbidden its use in environmental policymaking. This is becoming a bigger problem now that the cheapest environmental problems have already been tackled; the remaining pollution sources are likely to be the most costly to clean up.

The most glaring example of failure comes from the Superfund scheme, which sought to clean up America's toxic waste sites through tough federal laws. Though noble in its aims, it has proved hopelessly bureaucratic and even counterproductive in practice. The scheme often set the standards for risk so high (clean enough for toddlers to eat the dirt in a theoretical nursery located on the site, even if the dump was located in an industrial park) that many polluted sites simply never got cleaned up. Superfund has already sucked up so many billions of dollars—with much of that money going to line the pockets of lawyers—that some environmental experts are now calling it Stupidfund. Worse yet, the scheme is likely to cost America another $14 to $16 billion more over the next decade or so.

The old top-heavy approach to solving environmental problems is woefully inadequate to the task of dealing with tomorrow's thornier environmental problems. One reason is society's ever-rising expectations: the green goalposts keep moving. That's good, but squeezing the last 5 percent of a particular pollutant out of the air or water is usually much more expensive than it was to remove the first 5 or even 50 percent. That is when the sensible tools of market environmentalism come in handy.

Also, the world now has a much better understanding of mankind's impact on the environment—and that often brings to light less obvious problems. The link between an obscure chemical used in hair sprays and a hole in the ozone layer, for example, and the relationship between carbon dioxide and climate change are much harder problems to understand and tackle than the typical concerns of three decades ago, when Cleveland's polluted Cuyahoga River spontaneously caught on fire.

Thilo Bode, the former boss of Greenpeace International, says that today's greens face a dilemma:

> The ability of the market system and economic growth to destroy nature is undeniable. Since economic growth is however a prerequisite for overcoming poverty in economically emerging countries, the environmental movement is in the horns of a dilemma. To be in favor of growth undermines its legitimacy, but to be against growth opens the movement to the accusation of behaving like a colonial missionary for the environment. A strategic challenge to environmental associations is to close the dialectical gap between combating poverty in poor countries and conserving their natural life support bases. The issue has become even more pressing since 11 September. Forcing environmental standards on developing countries through WTO regulations amounts to ecological colonialism. The only

> solution can be for industrial nations to change their ways of doing business and to reduce resource consumption, and to stimulate economic growth only by reducing natural resource consumption. This is not a "pre-concession" but a moral duty of industrial nations that have gained prosperity by ruthlessly exploiting nature.
>
> The environmental movement must make central issues out of countering a host of ills: the market's failure to behave responsibly, destructive economic growth, the power of commercial interests, including profit-driven technical progress, and the incompetence of national and international governing systems. Can these issues gain support from the majority of the public? Can they attract new membership? In the end, no. Indeed, the ecological movement will be forced to pick an argument with most people, even though a small minority is committed to change. Whoever wants to fundamentally change society cannot expect that most people will immediately go along with them.

While Bode has put his finger on the problem, he despairs that the solution he proposes—that the rich world must become much more efficient in how it uses natural resources—will never be accepted by society. It is a pity that he gives up so readily, for market forces are tailor-made for his proposed solution: markets cannot do everything, but encouraging efficiency is one thing that they are extremely good at. Happily, many other greens are now turning to the market as the way out of their dilemma.

Indeed, a fresh wind is blowing through the world's environmental-policy realm. From Boston to Brussels to Beijing, governments are beginning to tinker with various types of innovative environmental policies, ranging from emissions trading to green taxes. These economic instruments could help them harness the power of market forces for the sake of the planet's health. They could channel the selfish profit motives of firms and individuals toward the common good by infusing environmental policies with economic incentives. This could have a dramatic impact on the way that the modern world thinks about the environment—and especially in how it tackles global warming, the environmental problem with the biggest potential price tag attached.

Market-based environmentalism differs from the conventional sort in that it tries to influence behavior by sending price signals, rather than merely through regulations spelling out desired pollution levels or mandated pollution-control technology. The great weakness of the conventional approach is that it gives firms little leeway in how they achieve preordained pollution targets, usually set in faraway federal capitals. These targets often fail to take into account local environmental conditions or the particular economic circumstances of the firm involved. Companies are compelled to adopt specified processes or even technologies, and to take on equal responsibility for meeting their pollution goals, regardless of their particular marginal cost of doing so. Such solutions stifle innovation. What's more, companies have an incentive to hide "least expensive fixes" to problems out of fear that revealing them will encourage more mandates—and, worse yet, to use political influence to win exemptions, delayed enforcement, and other special favors.

Robert Stavins, a Harvard economist, argues that market instruments can do precisely the opposite: since it always pays firms to clean up a bit more if a sufficiently low-cost process or technology can be identified, such policies can prove a powerful stimulus to innovation and environmentalism at once. Though there are dozens of variations on the theme, Stavins divides up market-based instruments into four broad categories: tradable permits, charge systems, government subsidy cuts, and reductions in what he calls "market frictions."

With typical tradable permit schemes, governments decide acceptable levels of pollution and allocate those limits without charge among companies in the form of tradable permits. Those firms that can cut pollution at the lowest cost will have spare credits to sell, while those with high abatement costs can buy from them. This is not just the fantasy of economists in ivory towers. The Montreal

Protocol (the plan for saving the ozone layer), for example, includes a provision for the trading of chlorofluorocarbons. Chile, which began its free-market reforms under the dictatorship of General Augusto Pinochet several decades ago, has successfully auctioned bus licenses and traded "particulate" pollution.

Even so, it was America that really showed the world the true power of trading systems—and economists at such market-minded outfits as Environmental Defense and Resources for the Future led the way. In a prescient article published in 1989, Stavins chronicled the first hesitant steps toward this more sensible approach to protecting nature:

> There was a time when the only serious consideration given to market-oriented environmental-protection policies was by economists (in academia, government and the private sector), but a new environmentalism that embraces these approaches is emerging. The Environmental Defense Fund [the old name for the group] was first among the major, national environmental organizations to advocate incentive-based policies; now there is growing interest in these approaches among other groups as well.

To understand how radical that strategy really was, we need to look to the presidency of the elder George Bush, when the environmental problem dominating the news in America was not global warming but acid rain.

Market Mechanisms Stumble over the White House Dog

The most striking environmental success story of the 1990s is probably America's trading system for sulfur dioxide (SO_2), the main precursor to acid rain. Environmentalists had been trying to get Congress to tackle that problem since the late 1970s, but regional politics had led to gridlock. Midwestern states like Illinois derived much more of their electricity from older coal-fired power plants than did northeastern states like New York, which generally had newer and less polluting power plants. Whenever anybody proposed tough across-the-board technology standards to control SO_2, the northeastern states squawked: Why should they be forced to spend massive sums on new equipment when they were not really a significant source of the acid rain problem? The midwestern states' main rebuttal was that fairness dictated that everybody should face the same new standards: if their utilities had to spend big bucks, so should everybody else's.

By the time George Bush took office, in 1989, Washington was primed for a fresh approach, and a couple of straws in the wind suggested that a market-based alternative might be ready for the big time. A report issued by the General Accounting Office, the independent watchdog agency of the Congress, confirmed that a handful of early emissions-trading projects set up by the EPA had already saved America some $5 billion in air-pollution-control costs at no detriment to the environment. Another auspicious sign came from Project 88, an influential bipartisan report put out by a committee of experts commissioned by Senators John Heinz and Tim Wirth and led by Harvard's Robert Stavins. The title of the group's final report speaks volumes: "Harnessing Market Forces to Protect the Environment—Initiatives for the New President." Given that Bush had promised on the campaign trail to take the environment very seriously, the timing could not have been better, though mainstream green sentiment remained highly skeptical.

Fred Krupp of Environmental Defense recalls the nervous energy of those days: "The new President came from a tradition that recognized stewardship of the earth as a fundamental value. In addition, the President and his close advisers had indicated an interest in tapping market forces in the service of environmental protection . . . although anticipating that we would be caught in a partisan cross fire during the ensuing legislative debate, we sensed an opportunity both to advance our distinctive ideas and to break the logjam that had blocked acid rain legislation for the better part of a decade. We moved forward by proposing a tough mandatory nationwide cap on SO_2 emissions, paired with full flexibility for individual power plants to choose among various technologies and processes to achieve their caps." The key to the scheme was the introduction of tradable rights, combined with the credible threat of costly fines and punishment for noncompliance.

Such an approach may sound like common sense now. Indeed, it has become pretty mainstream thinking among greens. Back in 1989, though, when Krupp and his colleagues were promoting the SO_2 plan, those same greens ostracized Environmental Defense. Krupp describes it this way: "Our staff faced withering criticism from various interest groups, some on Capitol Hill and even some EPA staff, for our increasingly visible advocacy with the Bush Administration of our proposal. During this period, we continually reminded ourselves that establishing an emissions-trading framework for power-plant SO_2 reductions would be instrumental, if not indispensable, to our longer-range goals with respect to greenhouse-gas pollution reduction."

In the end, the group's efforts were worth it: George Bush, who had been lobbied intensely by the energy industry not to act on SO_2, finally agreed to a plan that would amend the sacred Clean Air Act to incorporate emissions trading. Krupp was invited to join the President at the Rose Garden ceremony for the hugely controversial announcement. To avoid the throng of reporters in front of the White House as well as any stray energy executives and greens that might be lurking around, Krupp was rushed through the underground tunnels below the White House to the press conference. In the frenzy, he stumbled right over the house's most popular resident—Millie, George Bush's aging dog.

America's first pooch survived the unprovoked assault—though Krupp's colleagues still tease him about it to this day. What no one teases him about, though, is the sensibility of trading. Bush gave credit where credit was due: "I have no pride of authorship. Let me commend Project 88 and groups like the Environmental Defense

Fund for bringing creative solutions to long-standing problems, for not only breaking the mold, but helping to build a new one."

Experts at Resources for the Future, another group that deserves plaudits for championing emissions trading, estimate that SO_2 trading saved America $1 billion a year over the 1990s. What is more, emissions of SO_2 have fallen by more than predicted; in fact, the cuts made by industry have been even steeper than required by the law. While other unrelated factors also deserve some of the credit for those reductions, there is no denying that the SO_2 trading scheme merits praise.

Of course, the approach is not perfect: despite the fact that trading has reduced sulfur deposition, some places can still become localized "hot spots" of pollution that require further action. Also, a credible threat of punishment for cheaters and expensive real-time monitoring equipment are needed. That combination may not work for more complicated pollution problems, or in very poor countries. On the whole, however, America's SO_2 trading system has been such a success in both economic and environmental terms that many parts of the world are now copying it. If she were still alive, even Millie would probably forgive Fred Krupp.

America has led the way in emissions trading, but other countries have gone much further when it comes to getting the prices of goods and services to reflect their true environmental impacts ("getting prices right," in the jargon of environmental economists). Led by Scandinavia, European countries are embracing comprehensive reforms that shift taxes away from labor (by reducing social security levies or income taxes) to specific environmental harms. The guiding principle is that the polluter pays for the harm that his actions contribute to the environment.

The Organization for Economic Cooperation and Development (OECD), a quasi-governmental think tank for the world's rich countries, argues that there is growing evidence that such green taxes actually work, and that they make economic sense. Sweden's experience is telling. In 1991 the country introduced a tax on fuels based on their sulfur content: this led to a drop in sulfur content

50 percent beyond legal requirements, and it stimulated power plants to invest in new abatement technology. Norway's carbon tax, also levied in 1991, lowered emissions by over a fifth from power plants. Denmark's hefty tax on nonhazardous wastes led to sharp declines in waste from households and from construction. And as far back as the 1980s European countries phased leaded gasoline out of the market by pricing it higher than unleaded—i.e., by imposing a lead-content tax.

The main problem with green taxes is that they are too often blunted by blanket exemptions granted for heavy industry on such bogus pretenses as preserving "national competitiveness"—an example of corporate welfare that is often repeated in the environmental realm. Which brings us to the need for another sort of market-based reform: the reduction of environmentally harmful subsidies. In a sense, this reform is a mirror image of imposing a green tax: both aim to get prices right. Yet subsidies—lavished on everything from agricultural fertilizers to underpriced water to cheap electricity—are probably the single biggest distortion of the markets of the developed world.

The OECD reckons that if its member countries embraced a tax on carbon and on chemicals, and reduced un-green subsidies, they could improve their environment dramatically in two decades for a small price: less than 1 percent of the 2020 GDP, assuming the revenues are "recycled" back to taxpayers in some way. Now consider this: the governments of developing countries often do even greater harm to their environments through subsidies than do rich countries. They typically subsidize water and power in the name of the poor, but in fact urban elites end up sucking up that subsidy—leaving less money to help the genuinely impoverished living in slums and in the countryside. Such subsidies are nothing short of perverse, according to Norman Myers of Oxford University, as they do double damage: they distort market efficiency, and they encourage behavior that harms the environment. Though development banks say such subsidies add up to $700 billion per year, Myers estimates the global figure is over $2 trillion a year.

The numbers do not do justice to the harm done, however: EU countries subsidize their fishing fleets to the tune of just $1 billion a year, but that has spurred enough overfishing to drive many North Atlantic fishing grounds to near collapse.

The fourth market reform advocated by Stavins involves reducing "frictions" in existing markets and barriers to the creation of new ones. He suggests that if governments facilitate the voluntary exchange of rights, it will promote the more efficient allocation of scarce resources. He points to the example of emerging water markets in the American West, which has long been both thirsty and cursed by distorted federal policies on water. As recently as 1990, farmers in one part of California were paying as little as $10 for water to irrigate an acre of cotton, while residents of other parts of the state such as Los Angeles paid $600 for the same amount of water. As subsequent reforms have spurred the development of water markets, voluntary exchanges are now taking place throughout the West.

Well-functioning markets depend on good information flow, and polluting firms typically do not like to boast of their dirty deeds. Therefore, perhaps the simplest and most powerful application of market forces lies in requiring information disclosure. Governments can legitimately (and cheaply) boost markets by insisting on transparency in environmental reporting. Once properly equipped with information on the greenness of companies, consumers can vote with their pocketbooks. As long as eco-labels—which are similar to nutrition labels on food packages today—do not discriminate against imports, they need not be inconsistent with free trade or international laws.

European countries have had eco-labeling schemes for some time, most notably Germany's Eco-Angel program; America has Energy Star. In America, the Toxic Release Inventory requires firms to reveal their releases of more than 350 toxic chemicals. Using this once obscure data, "environmental justice" groups have created Internet sites where any ordinary person can plug in her postal code and find out how dirty the neighborhood firms really are.

The biggest bang for the buck may have come with Indonesia's Proper program. Local officials had scarce resources to enforce existing pollution laws by cracking down on violators, but with the help of the World Bank, they designed a five-tiered scheme that ranked companies by their environmental compliance: from gold for beyond compliance to black for flagrant violations. They publicly applauded the top few and gave the worst violators six months before their names would be made public. Astonishingly, most of those scofflaws rushed to invest in abatement technologies and otherwise clean up their act for fear of public scorn. The Proper scheme is now being adopted by the Philippines and several Latin American countries.

Another area where market reforms would help is in valuing environmental "goods" like biodiversity. Organizations like the Katoomba Group—which includes representatives of financial, forestry, and energy companies; environmental outfits; and development agencies—are enthusiastically trying to create markets for such goods. The idea, says Josh Bishop of the World Conservation Union, is to "speed the development of markets for some of forestry's ignored 'co-benefits,' like carbon storage, biodiversity conservation, and watershed protection. That can produce new revenue flows for forest owners and create an incentive for more sustainable management."

This approach shows promise. When federal environmental regulations enacted in 1989 forced New York City to clean up its water, officials initially thought they needed to build a new filtration plant costing more than $6 billion. However, thanks to the eco-entrepreneurship of a few officials, the city did not build that plant. Instead, it pays a fraction of that amount to protect the watershed in upstate New York—the source of the city's drinking water—and to encourage communities to manage their forests and agricultural land better. As similar experiences in Costa Rica and elsewhere have shown, paying for nature upstream can be cheaper than cleaning up water downstream after it has been fouled.

All very enlightened, you might think, but individual watersheds and forests are small potatoes compared to the entire global energy economy. The value of the former is usually measured in billions of dollars, after all, while the latter runs into the trillions. What happens when the stakes are really high—as in the global warming debate? Climate change has the potential to overturn all of the ground rules of the global economy. And it poses a direct threat to oil majors, which have historically been among the most powerful and reactionary forces in society. Lord Browne, the head of BP, even describes the challenge in philosophical terms: "Climate change is an issue which raises fundamental questions about the relationship between companies and society as a whole; and between one generation and the next. It is an issue which is about leadership as well as science."

Perhaps it is asking too much of the market to solve this problem. Even so, there are already promising signs that market-based approaches can at least help bridge the gulf between the increasingly irresistible pressure to tackle climate change and the seemingly immovable obstructionism of the world's fossil-fuel giants.

Climate Change Meets Big Oil

"Dear Tony," starts the memo from one earnest policy wonk to another. "Knowing our shared interest in global climate change, I have enclosed a report on the benefits of greenhouse-gas emissions trading . . . This new report by the nonprofit Pew Center on Global Climate Change supports the view that a flexible, well-designed trading system will significantly reduce the costs of climate change mitigation . . . With best regards, Bill."

These are not the nerdy mumblings of a bureaucrat or scholar. The letter was actually sent by Bill Clinton, when he was still President of the United States, to Tony Blair. It demonstrates the inroads that market-based approaches like emissions trading have made into mainstream thinking about the environment. It also reveals just how influential the Pew Center has become. In fact, it is not an exaggeration to say that this group has helped change the terms of debate on global warming in America.

As recently as 1997, in the run-up to the signing of the Kyoto Protocol, skeptics dominated the public debate in America on climate change. Leading the opposition was the Global Climate Coalition (GCC), an umbrella group representing heavy industry, especially the fossil-fuel and railway sectors. Through public campaigns and intense lobbying, GCC executives sought not merely to sink the Kyoto treaty but to persuade the public that the very notion of climate change was a hoax. That was a sharp contrast with the less hostile attitudes of big business in Europe and Japan. Even when faced with new scientific evidence of the human role in climate change, the GCC remained unbowed.

When Eileen Claussen, head of the Pew Center, started to challenge the GCC publicly after the Kyoto conference, however, the picture began to change. Claussen, a former government bureaucrat with experience in dealing with international environmental problems (including the hole in the ozone layer), approached leading industrial firms she thought might be open to a different strategy. She was sure that big business was not uniformly opposed to action. She was right. A handful of companies agreed to join her under Pew's umbrella. Giants such as United Technologies, Intel, American Electric Power (AEP), and a number of other Fortune 100 firms now openly accept that the evidence for climate change is sufficient to take seriously. Many have conducted audits of the amount of greenhouse gases emitted by their plants and are taking on voluntary targets for emissions cuts. Some firms even think reducing emissions could be profitable.

By spending pots of money on ads, brochures, independent reports, and high-profile conferences, the Pew Center has managed to earn credibility as an honest broker. The Pew's media skills meant that companies found strength in numbers and a powerful voice. Today, mass defections have turned the GCC into a spent force, while Pew's blue-chip member companies have annual sales of well over $500 billion. Where moderate firms once kept silent, they now have a forum. "In the past, only the bad guys were out there in the media," says Claussen. "We raised the ante by being so public."

DuPont, the world's largest chemical firm, promised to slash its emissions of greenhouse gases by nearly two-thirds from 1990 levels before 2010, while holding total energy use flat and using renewable resources for one-tenth of its energy worldwide. The board of Royal Dutch/Shell, the oil giant, decided that all big projects must take into account the likely future cost of CO_2 emissions. Shell's Aidan Murphy explained the logic behind his firm's use of various, slightly speculative projected "carbon prices": "We know that $5 and $20 are surely the wrong price, but everyone else who assumes a carbon price of zero in future will be more wrong. This is not altruism. We see it as giving us a competitive edge."

Frank Loy, a top American negotiator on climate change during the Clinton presidency, says that many big firms have "shifted from being climate skeptics to climate activists." Indeed, a number of them openly defied the Bush Administration in 2003 by supporting a domestic "cap and trade" bill introduced by Senators Joe Lieberman and John McCain. Loy thinks that at least some of the credit for this U-turn must go to the Pew Center. It created the public space for good corporate citizens to emerge in this highly polarized, and often dishonest, debate. Even so, it was not easy for the early joiners, who were berated by their peers for breaking rank. No one knows this better than BP's boss.

The Greening of Browne

Lord Browne makes an unlikely Don Quixote. He transformed British Petroleum, a laggard in the world league table of oil majors, into one of the world's biggest oil companies through breathtakingly swift acquisitions of America's Amoco and Arco in the late 1990s.

Yet the man who made a name for himself as a ruthless cost-cutter and daring deal-maker may have an idealistic streak too. He has invested heavily in renewable energy; BP's Solarex division is the world's biggest solar equipment manufacturer. He has started a hydrogen division to prepare the firm for the arrival of fuel-cell technology. He has also been putting solar panels on top of his conventional gas stations across the world, and cleaning up the nastier pollutants in his gasoline. Most surprising, though, have been his actions on climate change.

He was the first oil boss to break ranks with the GCC in 1997. He publicly accepted that climate change was real, and he declared his support for the aims, if not all the devilish details, of the Kyoto process. He also committed his firm to reducing its emissions of greenhouse gases by 10 percent below their 1990 level by 2010—well ahead of any legal requirements. Initially, many environmentalists were deeply skeptical, calling his move a "greenwash." They noted that BP's internal emissions are trivial in comparison to those released when consumers guzzle its gasoline. Also, some suggested his actions were merely designed to kill proposals for a hefty carbon tax on gasoline—which could have hit the bottom line of oil companies pretty hard. His fellow oilmen considered him either a naïve turncoat or a sophisticated fraud. Exxon's boss, Lee Raymond, scoffs at BP's green investments: "Show me the money!" Yet the BP boss says he expected such hostility: his senior managers warned him, he says, of a "first-mover disadvantage" on this issue.

So why then did he set out on the green path? One reason, he explains, was the science: both his researchers and he himself (a scientist by training) had become troubled by the growing, though still inconclusive, evidence of global warming. "A break was inevitable because companies composed of highly skilled and trained people can't live in denial of mounting evidence gathered by hundreds of the most reputable scientists in the world. Our people, in common with everyone else in the world, have hopes and fears for themselves and their families."

He says he was concerned too about the implications for his business: "We simply cannot survive for long if we remain so out of tune with our consumers' perceptions, and the next generation's attitudes." Another factor was a conviction that oil companies "must engage in the debate, and not be shut out as the bad guys." With his bold stance, Lord Browne may be doing nothing less than preparing BP for the unthinkable: life after oil.

Revealingly, he chose to fulfill his pledge to cut greenhouse-gas emissions by turning to market forces. With the help of Environmental Defense, BP launched an innovative internal market for emissions trading among its many divisions around the world. Even some experts thought it wouldn't work. In 2002 the firm announced spectacular results: it had met its target for emissions reductions seven years ahead of schedule. It did so through a combination of efficiency, technology, and better management of energy—all of which happened only because there were clear orders from the top that tackling greenhouse gases was a priority.

Most impressive of all is the fact that the company achieved this green goal—which is similar to the Kyoto targets, derided by the GCC as ruinously expensive—essentially for free. He explains: "We've met it at no net economic cost—because the savings from reduced energy inputs and increased efficiency have outweighed all the expenditure involved. That's a particularly noteworthy point, a positive surprise, because it begins to answer the fears expressed by those who believed that the costs of taking precautionary action would be huge and unsustainable."

The lesson for the world at large is clear. If those at the top send an unambiguous signal that climate change is important, and if they encourage market-based solutions, industry will respond with a blunderbuss of innovation that produces the lowest-cost solution to this thorniest of problems. After all, as the next three chapters (which describe the good, the bad, and the ugly of energy technology: fuel cells, oil drilling, and nuclear power) make clear, government policies have an enormous influence on the pace and direction of innovation in this field.

Lord Browne sums up his firm's experience this way: the costs of tackling climate change are "clearly lower than many feared. This is a manageable problem." Coming from the boss of an oil giant, which has more to lose than most, those words add up to a compelling case for action. Will governments listen?

ENERGY TECHNOLOGY

Bigger than the Internet

8

The Future of Fuel Cells

THE THUNDERING TOM-TOMS seemed to herald something big. Several dozen people from around the world recently descended upon an idyllic country retreat in Canada for an energetic powwow about the future of our planet. They sat in a giant circle with native drums of every imaginable size and shape, got in touch with primal rhythms, and proceeded to bang away like mad until green inspiration struck. They then strategized about how to move the energy world beyond the dirty but durable workhorses of today—fossil fuels and internal combustion engines. They divined that the future belongs to a magical technology that would combine hydrogen, the most common element in the universe, with oxygen, freely available from air, to produce squeaky-clean energy.

Here's the weird part: those peculiar percussionists were not New Age environmentalists, but senior executives from the world's biggest car companies, energy firms, and research laboratories. Indeed, the Hydrogen Interactive was organized by a new division of BP, the oil giant, as a brainstorming session on the future of energy. The gurus agreed that the future belongs to fuel cells.

So what are fuel cells? In a nutshell, they are big batteries that run for as long as fuel is supplied. There are various types, but nearly all work by combining hydrogen with oxygen to produce electricity, while resultant emissions are no worse than water and heat. If the hydrogen is extracted from some intermediate source, say a hydrocarbon fuel like gasoline, then there will be some dirty emissions (though still less than if that same gasoline had been burned in a conventional combustion engine). But if the fuel supplied is pure hydrogen made from renewable energy, then this power source will release no greenhouse gases or local pollutants whatsoever—the ultimate environmentalist dream.

One especially promising type of fuel cell is the "proton-exchange membrane," or PEM. This is a sandwich of two electrodes—an anode and a cathode—with a solid polymer membrane stuck in the middle. At the anode, a platinum catalyst embedded in the fuel cell stimulates the hydrogen to give up its electron. While the hydrogen protons slip through the tiny pores in the membrane, the electrons are forced to travel around an external circuit, producing a current that can power a home, a computer, or a car. When the protons reach the cathode, they join with the electrons and combine with oxygen from the air to create water and heat. That is why, if the fuel supplied is pure hydrogen, the result is clean energy with no harmful emissions.

While fuel cells may be the technology of the future, the basic concept is actually more than 150 years old. Long before the development of the internal combustion engine, scientists and inventors in Europe were fiddling with ways of combining oxygen and hydrogen to produce electric current. Though earlier scientists had gleaned clues about the "fuel-cell effect," it was Switzerland's Christian Schoenbein, now best remembered for discovering ozone, who first published a clear explanation for it. It was his friend, an English judge and tinkerer named William Grove, who turned that insight into a workable product known then as a "gas battery"—an early version of today's fuel cell.

Promising as these discoveries were, neither these two men nor the many that followed them could find a way to turn fuel cells into a commercial product. The laboratory apparatus they came up with was too clunky, complicated, and ultimately too costly to be practical. Fuel-cell research fell into obscurity for decades, and Grove and Schoenbein are long forgotten. But maybe not for much longer. Ulf Bossel, who is a direct descendent of Schoenbein, managed the fuel-cell development program at the energy giant ABB: "When Christian Friedrich Schoenbein and William Robert Grove made their discoveries, Europe still conducted its evening activities by the light of a candle, and most of the United States was still the property of Indian tribes. But the names Schoenbein and Grove will undoubtedly become more recognizable in the next decade or so; perhaps much as the names Otto and Diesel are recognized today, even though their accomplishments [the development of, respectively, the four-stroke gasoline engine and the diesel engine] took place more than a century ago."

As the enthusiasm of the Hydrogen Interactive made clear, though, nostalgic relatives aren't the only ones who believe that the fathers of the fuel cell are about to hit the big time. Why the sudden buzz about an obscure technology from the days of the horse and buggy? The truth is, this technology has been making quiet but steady strides toward commercialization for decades. Research was renewed when America's space program began a search for reliable power sources for its astronauts. Conventional batteries were bulky and could not last the necessary two weeks of flight time. Solar power would not do either, as space travel requires astronauts to pass regularly through the earth's shadow.

The advantage of fuel cells was the fact that their power "density" was about eight times that of the best batteries available at the time—meaning less weight and volume to lug into space. Since NASA could afford the prohibitive cost of the technology back then, it placed orders for fuel cells and even bankrolled more research. From the Gemini program to the Apollo lunar missions to today's space shuttles, fuel cells have been an indispensable part of NASA's success. One additional benefit is that the astronauts' thirst can be quenched by the fresh, sterile water that is the only exhaust produced by these successors to Grove's magical "gas battery."

Even as fuel cells lifted off into space, however, they failed to take off back home. Researchers at universities around the world, as well as engineering giants such as United Technologies, Siemens, and General Electric, experimented with a variety of fuel-cell technologies—solid oxide, phosphoric acid, alkaline, solid polymer, direct methanol. By the 1980s, however, they had little to show for their work. Fuel cells had made few inroads into the commercial market, and even enthusiasts began to worry that they would always remain too costly to be practical.

Despite decades of frustration and failure, fuel-cell fans are once again jumping for joy. The collision of environmentalism, market liberalization, and technological breakthroughs is finally bringing the cost of fuel cells down to earth. The first push came from politicians. Carmakers began to take fuel cells seriously after California's controversial Zero Emission Vehicle mandate decreed that a tenth of all cars sold in the state by 2004 must produce no local emissions at all—or else the car company could be banned from selling any cars in the state. Authorities imposing the blunt tool of the ZEV mandate initially thought that battery-powered cars would be the magic bullet to achieve zero emissions. But as my ridiculous experience renting an electric car in Los Angeles showed, the all-knowing regulators picked a dud: batteries are simply not good enough to crack the mass market for cars.

Confronted by that failure, California's regulators agreed to apply that ZEV mandate a bit more flexibly—easing up on targets and timetables, but still maintaining the hard edge of a potential ban. The happy result was that the California edict proved a big boon to fuel-cell research. Ferdinand Panik, a former DaimlerChrysler official who is a top scientist in this field, says quite bluntly that he would not have received the hundreds of millions of dollars in funding for his firm's fuel-cell program in the 1990s if it were not for California's regulations. Car firms and fuel-cell specialists, such as Canada's Ballard Power Systems, have spent over $2 billion on research and development in this area already—and are increasing their commitments as fuel cells get cheaper and more efficient.

Enthusiasts argue that the electricity industry will be turned on its head by fuel-cell "micropower" units that are already starting to trickle onto the market at places like New York City's Condé Nast skyscraper. Given the inadequacies of today's battery technology for such products as laptops and mobile phones, they postulate that really tiny fuel cells will transform the market for portable power too. The big game, though, continues to be transportation—hydrogen's promised land.

Enter the Hypercar

For the last decade or so, a crack team of engineers, car designers, and software experts has been busy developing the power plant of the future. Amory Lovins, the head of the Rocky Mountain Institute and the mastermind behind this initiative, describes the resultant "Hypercar" in his usual understated manner. They have, he says,

> designed an uncompromised, manufacturable, production-costed, midsized SUV concept car. It's as roomy, comfortable, and sporty as a Lexus RX-300 or a Ford Explorer; is as safe even if it hits one, although they're twice its weight; and should cost about the same to make. Yet at [a] 99-mile-per-gallon equivalent, it uses 80–82% less fuel and no petroleum, and emits only hot water. Instead, it can drive 330 miles on 7.5 pounds of safely stored compressed hydrogen, or about 600 miles on 14 pounds using the latest tanks, because of the fuel cell's doubled efficiency and the car's lightness and low drag. Such Hypercars could enter early volume production around 2004–2005 and transform automaking, the world's largest industry, within two decades.

> For the United States, such vehicles of all shapes and sizes could ultimately save eight million barrels of crude oil per day—like finding an inexhaustible Saudi Arabia by drilling in the "Detroit Formation." A global Hypercar fleet could save as much oil as OPEC now sells. Think negamissions in the Gulf—Mission Unneccessary.
>
> Hypercars' quintupled-efficiency, highly integrated, radically simplified, software-rich design makes the car ready for hydrogen, with fuel cells small enough to afford and hydrogen tanks small enough to fit. Such cars are the key to achieving a climate-safe hydrogen economy profitably at each step, starting now.

Phew! Lovins's mind whirrs at such breakneck speed that it is hard to keep up with him. Setting aside the blizzard of technical details for the moment, it's pretty plain that he's bullish on the future for fuel cells.

Is Lovins nuts? If so, he's hardly alone. In 1997 Bill Clinton gushed that "Ballard Power and United Technologies are leading pioneers in developing fuel cells that are so clean . . . their only exhaust is distilled water." Senator Tom Harkin, a longtime believer in this technology, was even more effusive: "Fuel cells can be made in any size to fit everything from pocket-held devices to large power plants. They are perfect for a dispersed and robust energy infrastructure." Romano Prodi, the head of the European Commission, insists that he wants to be remembered for only two things: the expansion of the European Union eastward and hydrogen energy. The most meaningful endorsement comes from Bill Ford, the great-grandson of the car firm's famous founder and its current chairman. Though he has a vast fossil-fuel empire to defend, he recently declared that "fuel cells will finally end the 100-year reign of the internal combustion engine."

Despite such enthusiasm, there are still hurdles that could yet trip up this technology as it races from the laboratory toward your office, garage, or shirt pocket.

The Vanguard of the Electric Revolution

What do a brewery in Japan, a credit-card processing center in Nebraska, and a police station in New York's Central Park have in common? They all get their electricity from a micropower plant located on-site, powered by a stack of fuel cells. Each chose fuel cells over conventional power plants for a different reason, which suggests that this technology may have legs. Japan's Asahi brewery chose fuel cells because they can operate using hydrogen from the waste methane gas produced by the firm's industrial processes. The First National Bank of Omaha, which handles backroom operations (like tracking purchases) for credit-card issuers, chose fuel cells after a power outage in 1997 cost the firm a fortune. Fuel-cell systems can provide high-quality electricity with 99.9999 percent reliability, while the American electricity grid averages about 99.99 percent—a big difference if your business depends on computer systems. The police in New York were influenced by cost and reliability, of course, but they saw yet another compelling advantage in fuel cells: they were simply much cleaner than the other choices available, such as diesel generators, and therefore better suited for placement in a city park.

These early adopters had good reasons not to pick a power plant based on the purchase price alone. In the real world, though, most consumers do make their decisions based chiefly on economics. Will fuel cells once again flounder in the face of competition, as they have since Grove's day? Not necessarily. Research into PEM fuel cells has greatly reduced the amount of platinum needed, and has made the electronics cheaper. A decade ago, the amount of platinum required by a stack of PEM fuel cells to run a car cost $30,000; less than a few hundred dollars' worth is required now.

The cost of a fuel cell has already fallen to below a few thousand dollars per kilowatt of generating capacity, but car firms still have a lot to do: to compete against the internal combustion engine, fuel-cell costs must come down to about $50 to $100 per kilowatt. Carmakers think mass production will help them close the gap soon, and big power-generation companies hope to ride their technology coattails. Market research conducted a few years ago by Arthur D. Little, a technology consulting firm, suggested that consumers would spend $1,000 per kilowatt for the benefits offered by small combined-heat-and-power units. Today's incumbent technologies, such as coal generation, also cost a similar amount, so fuel cells have a fighting chance.

Does the notion of a fuel cell in your backyard anytime soon still sound far-fetched? Consider this pitch found on the website of one firm hoping to be a pioneer in this market:

> *Imagine your own reliable supply of electricity in a compact, quiet, self-contained package*—a fuel cell called the HomeGen. This new energy system, now under development, will generate electricity at your home. Because it is fueled by either natural gas or LPG, it will be both efficient and environmentally friendly. Installed in your back yard, the HomeGen fuel cell is being designed to provide 100% of your home's energy needs . . . The HomeGen fuel cell will be one of many distributed energy products marketed . . . [and] available for retail purchase through our global network of distributors.

Those confident promises come not from some fly-by-night technology upstart: they are from General Electric, one of the most admired corporations in the world. GE's Power Systems division is working with America's Plug Power, a pioneering fuel-cell firm, to develop PEM cells the size of a washing machine for the average consumer.

Other industry powerhouses such as ABB, Siemens, and United Technologies are also scrambling to develop fuel-cell technology. A number of firms think that fuel cells will take off in stationary applications before they hit the road under the hoods of cars. Together with Tokyo Gas, Ballard Power Systems, the industry's leader, has developed a little power plant that will let the Japanese gas utility's customers produce both electricity and hot water from natural gas—and so bypass the electricity monopoly altogether. Ballard has also teamed up with Coleman, a big camping and outdoors-goods firm, to develop portable generators based on fuel-cell technology.

If the durability of fuel cells improves by as much as boosters claim, then they could account for as much as one-tenth of the $50 billion-a-year global market for power-generation equipment by 2010. The proliferation of electronic devices in houses and offices would benefit from the high-quality power promised by micro-power generators. You don't have to be a gadget junkie or Silicon Valley trendy to benefit either. More than half of all American households own a personal computer (in Britain and Germany the figure is greater than 40 percent, and in Japan almost as high), and the number of households with multiple devices like fax machines, digital answering machines, sophisticated game systems, and the like is soaring. The popularity of gadgets suggests another potentially huge market for fuel cells: minuscule power.

Another Vodka for Your Laptop, Sir?

Picture a platoon entering hostile terrain. The location is too far from supply lines for bulky sensors or power-guzzling communications equipment to warn of an attack. Exposed, uncertain of enemy movements, the soldiers scatter handfuls of micro-sensors hither and yon throughout the danger zone to sense sound, motion, body heat, human smell, even the metal of the enemy's weaponry. The devices contain communications chips that warn of danger by beaming low-powered optical or radio signals to a portable control unit at the camp. Best of all, the coin-sized detectors are so cheap that the platoon simply leaves them behind when it moves on.

This scene is not as fanciful as it sounds. Even during the Vietnam War, American armed forces were scattering portable sensors along jungle trails, though at huge cost. As the know-how for micro-manufacturing such devices moves from the laboratory to the shop floor, sensor/transmitters will become cheap enough to scatter and forget, like so many spent cartridge shells. But the same problem that plagued the Pentagon's micro-sniffers three decades ago remains: the batteries needed to power them are far too bulky.

The trouble is that battery systems are pushing the upper limits of specific energy—the number of watt-hours they can store for a given weight. The best that conventional batteries can achieve theoretically is 300 watt-hours per kilogram (Wh/kg), though most manage barely half that in practice. That is nowhere near enough for the armed forces. The Pentagon has said that it wants to deploy portable equipment loaded with energy-guzzling features that would require up to 3100 Wh/kg by 2006. The physical properties of batteries make it impossible for them ever to achieve such goals. That's where a wholly different form of portable power comes in: micro–fuel cells.

The armed forces are not the only customer for a battery replacement, but their deep pockets certainly help pay for the early, expensive prototypes. The next generation of mobile phones, laptop computers, and personal digital assistants will all come with wireless networking circuitry built into them, demanding ever greater amounts of portable power. Samsung, a South Korean electronics firm, says its gadgets will need power sources capable of at least 500 Wh/kg within a few years. Experts believe that even today's most advanced lithium-polymer batteries could fall far short of future demand. That market opportunity has unleashed a breakneck global race to find replacements for today's lightweight batteries, and the best hope so far looks like a miniaturized version of the fuel cell.

Can PEM fuel cells, which weigh 100 kilograms or more when used in prototype electric cars, be shrunk a thousandfold to the size of a flashlight battery? Researchers are finding reasons to be optimistic. Fuel cells sipping methanol or hydrogen have far higher energy densities than batteries, they can run for longer periods before needing to be refueled, and refueling takes only seconds whereas rechargeable batteries need an hour or more. Moreover, users can simply replace spent fuel cartridges with new ones in much the same way that low-powered dry batteries can be swapped when exhausted.

The portable-power business faces none of the tricky questions that plague fuel-cell carmakers, such as where and how drivers will refill their fuel tanks with hydrogen or methanol. Laptop users in the future will be refilling their fuel-cell energy packs using capsules of methanol bought at kiosks or newsstands.

An even better argument for the miniature fuel cell, however, is cost. Though consumers pay only tens of dollars for them, today's rechargeable batteries are actually exorbitantly expensive in terms of bang for the buck: they cost $10,000 or more per kilowatt, compared with just $50 per kilowatt for a typical gasoline engine. So, as a replacement for rechargeable batteries, fuel cells do not have to meet an impossibly tough cost criterion. As Jerry Hallmark of the Motorola electronics firm put it recently: "We can afford to pay a few dollars for just a few square centimeters of fuel cells, but the car companies [who need much bigger fuel cells at much lower cost] cannot."

While the dozen or so groups seeking to make miniature fuel cells each have their own proprietary approach, they fall into two distinct camps. One wants to build on technology that has been demonstrated to work well in electric cars—the classical PEM fuel cell that is fed directly with hydrogen fuel—and then try to miniaturize it. The other side takes a radically different approach that has yet to work on the scale of a car engine. This involves feeding a PEM fuel cell directly with methanol fuel—that is, without first "reforming" it to release its hydrogen.

The first approach is simple and works well already in cars and stationary power plants. Jesse Wainright and his colleagues at Case Western Reserve University are trying "micro-manufacturing"—a technique for making very small devices on silicon wafers using techniques pioneered by the semiconductor industry. The beauty of the process is that dozens, even hundreds, of identical devices can be made simultaneously on a single wafer of silicon. When finished, they are diced up into individual chips and packaged. Sounds great, but as Wainright and others pursuing this strategy have discovered, that is easier said than done.

Most of the work on portable-power supplies in recent years has focused on the rival DMFC (direct methanol fuel cell) approach. Methanol cannot be used directly as fuel for automotive fuel cells, at least not with today's technology. Normally, its chemical structure first has to be reformed in a bulky, high-temperature unit that separates the hydrogen from the carbon in the fuel. A breakthrough came in the early 1990s, when a team at the California Institute of Technology's Jet Propulsion Laboratory (building on the work of many other pioneers) managed to produce electricity by feeding methanol directly to a small PEM cell. This advance, if translated into a workable product, could prove the key to cracking the portable-power problem. Methanol has several advantages over hydrogen: since it is a liquid at room temperature, it is easier to handle, and it also has a much higher energy density than hydrogen.

Unfortunately, DMFCs have problems of their own, ranging from membranes that are too leaky to catalysts that are too expensive. Government laboratories like JPL and Los Alamos and their counterparts in Japan and Europe, as well as private-sector firms like Japan's Toshiba and Korea's Samsung, are optimistic that they can soon crack these remaining problems. There is the slightest possibility that these heavyweights will be beaten to the punch by upstarts. Albany's Mechanical Technology, for example, claimed a big breakthrough in DMFC design in 2003 that it claimed would lead quickly to a commercial product—though it refused to reveal details. Medis, an Israeli-American firm, says it has come up with an entirely new approach that solves the "crossover" problem that bedevils the DMFC by using a proprietary liquid concoction, rather than solid material, as the "membrane" in its cell.

As a result, the firm is able to use a much higher concentration of methanol fuel, which boosts the system's power output. This approach also eliminates the need for complex fuel-mixing chambers and "micro-channels" to spoon-feed the fuel, as in rival DMFC approaches. Instead, the firm's fuel mixture goes directly into the fuel cell. Robert Lifton, the chairman of Medis, says that his company's fuel cells are so robust and flexible they can run on any alcohol. "We've got a lot of Russian researchers," he boasted at the cell's public unveiling in mid-2001, "and they've even got it running on vodka!" General Dynamics, a big American defense contractor, was pleased enough with the technology to enter into joint ventures with Medis.

It will probably be a couple of years before a viable product becomes available for portable power. Some companies are taking interim steps before tackling the ruthless commercial market head-on. Manhattan Scientifics of New York decided to launch its DMFC initially in the form of a portable battery recharger rather than as a battery replacement. The firm has designed its fuel-cell unit to be worn on the hip: while the phone is idle, its conventional battery will be recharged by the fuel cell. Motorola also planned such a hybrid as its first step. Even Medis, which boasts a much bigger power output, planned to introduce its fuel cells as a hybrid "fast recharger" first, and only later develop a full-fledged replacement for batteries.

The real test for this technology is consumer acceptance. Ordinary users have never seen such "energetic" devices before, except perhaps under the hoods of their cars. Unlike batteries, DMFCs are sensitive to temperature; they must breathe air for survival; and they exhale carbon dioxide and water vapor. Before consumers rush to put such a device in their pockets, they are sure to demand that fuel cells beat batteries hands down on cost and performance, if not on safety.

Noting that fuel cells emit a tiny bit of water vapor, one skeptical Sony executive quipped, "I don't want my mobile phone to wet my pants if I put it in my pocket." Any fuel-cell unit for a mobile phone must also work safely at temperatures inside a pocket or handbag. He pointed out that the aviation authorities in many countries have banned methanol and hydrogen on board commercial aircraft. Unless such regulations are changed—and lobbyists are hopeful, despite the clampdown following September 11—Sony worries that fuel cells may not take off in laptop computers.

In early 2002 the team at JPL came up with a new twist in DMFC design that its publicity department heralded as bad news for conventional batteries: "Researchers have reached an important milestone in developing a portable energy source that may someday give that hot pink, shades-wearing, drum-beating bunny a run for its money." Perhaps. But when I spoke with the team's leader, S. R. Narayanan, some months earlier, he recommended taking such proclamations with a grain of salt. He summed up the challenge for portable fuel cells eloquently: "Let's remember that batteries are nearly perfect in every way, except that they have to be recharged—and that is a very high hurdle indeed."

That hurdle must seem like a cakewalk to those long-suffering fuel-cell researchers aiming at an even tougher target: dislodging the internal combustion engine. In an article that appeared in *The McKinsey Quarterly* in mid-2002, Lance Ealey and Glenn Mercer argued that this could take many decades, if it happens at all: "Today's internal-combustion engine is far more advanced and efficient than its predecessors. Over the past 20 years, automakers have significantly improved its power, its fuel efficiency, and its emissions, with more changes to come . . . given the current economics of the internal-combustion engine, we predict that it will still be installed in 90 percent of all new vehicles sold in developed economies in 2015 and remain dominant in new vehicles for at least another decade after that, both as a stand-alone technology and as an integral part of hybrids." To see why the road ahead for fuel cells could be so challenging, consider the tale of the Baby Benz that road-tripped across North America.

Death on the Donner Pass

On June 4, 2002, a typical summer day in Washington, D.C., something not at all typical happened. Senators Carl Levin and

George Voinovich, midwesterners who were co-chairmen of the Senate's automotive caucus, cheered the arrival of an extraordinary German visitor who had traveled from California to the Capitol building. On that day, a specially equipped Mercedes-Benz "A-Class" sedan (a subcompact model not sold in America) became the first fuel-cell car to make the 3,000-mile-plus journey across the North American continent.

Fuel-cell fans everywhere rejoiced. The senators and other dignitaries lavished praise on fuel cells as the automotive technology of the future. Officials from DaimlerChrysler, the German auto giant behind the experiment, boasted that this latest version of their New Electric Car (NECAR 5) had created a huge stir along the route: "In some of the bigger cities like Chicago and Detroit people followed us to rest stops to ask us questions and start discussions." Perhaps imagining themselves back on the autobahns of their homeland, they also got the clean machine racing up to more than 90 miles per hour—an achievement that should put to rest any jokes about fuel-cell cars being merely fancy golf carts for old geezers. Ferdinand Panik, then Daimler's chief fuel-cell expert, was especially pleased: "It's a big, big step for a new idea, trying to look beyond the capability of fossil fuels."

Without doubt, the NECAR 5 achieved a milestone in automotive history. However, scratch the surface, and details emerge that suggest that the road ahead may still be quite bumpy for fuel cells in automotive applications. *The Hydrogen & Fuel Cell Letter*, an industry publication, did a lovely story on the cross-country trip that explains how nerve-racking the whole journey really was—and how very close to failure the Baby Benz came:

> Big Trouble struck on the first day when the DaimlerChrysler caravan—two SUVs and a van accompanying the fuel-cell car—hit snow on the Donner Pass, the legendary Sierra Nevada crossing where hardy early pioneers came to mortal grief a century and a half ago. Like the ill-fated members of the Donner Party of 1846, snow laid low NECAR 5. "We had a huge problem," related [Daimler of-

ficial Wolfgang] Weiss. "We didn't have a clue what the trouble was."

Water from melted snow had shut down a vital electronic component, and the car of the future had to be towed, rather ignominiously, to a dealer nearby. Engineers puzzled over it for hours before they figured out the problem and sent the car on its way once again.

Another unflattering detail that the headlines did not convey was the elaborate fuel infrastructure the car needed. After all, you can't just pull up to your corner Exxon or Shell station and fill up on methanol or hydrogen fuel. Actually, you could do so if you lived near the Munich airport, which maintains a hydrogen refueling station, or in Sacramento, California, which happens to be home to the California Fuel Cell Partnership, a consortium of interested companies and government agencies that runs a refueling station for such cars. Daimler officials had to organize refueling trucks every 300 miles or so to ensure that their baby got its bottles of nourishment. With so many logistical and technical complications, the company did not even put out the first press release until the car had reached Michigan!

The world is just not ready to deliver hydrogen fuel at every street corner, the way gasoline is distributed today. Some people think that hydrogen infrastructure will be far too costly—perhaps $100 billion in the U.S. alone—to build for decades to come, and they propose using some interim fuel during the transition. Many big oil companies, which have massive sunk investments in the form of gasoline distribution infrastructure, initially endorsed this view. But others reckon that the investment needed will be much less, and they insist that going directly to hydrogen is the only sensible option. Many car companies, for whom hydrogen is a much easier fuel to feed into a car, are in this camp, as are nearly all environmentalists.

Few dispute that the greenest and most elegant way to feed fuel cells is using hydrogen fuel directly. In the future, the hydrogen may even be derived from renewable energy. Some researchers are investigating whether the hydrogen released by decaying plants can be captured for use, while others are looking into wind or solar farms to extract hydrogen by splitting water atoms (electrolysis). Until this utopia arrives, however, some firms are thinking about using interim fuels such as methanol or gasoline to extract hydrogen on board the vehicle using a small chemical-processing unit called a reformer. Such cars would not be emission-free, but they would still be cleaner than today's vehicles.

Methanol champions argue that their fuel, unlike gasoline, can be produced easily from a variety of sources, ranging from natural gas to "biomass" (plant matter, cow dung, and such like). This could make methanol attractive in poor countries, which use a lot of biomass. Rich countries may also prefer methanol because it would reduce their dependence on OPEC.

The best argument for reforming methanol is that it clearly works, while gasoline reformation remains in doubt. Gasoline is a far more complex fuel than methanol, and it contains carbon-carbon molecular bonds that take a lot of energy to break apart. That is why gasoline reformers must operate at high temperatures (i.e., 800–900 degrees Celsius), while methanol reformers run at perhaps one-third that temperature. The methanol backers gloat that they have more or less solved the chief technical puzzles: the NECAR 5 that traveled from California to Washington boasted a reformer that powers a 75-kilowatt fuel-cell stack, while gasoline reformation remains bedeviled by technological problems.

Researchers led by a consortium including ExxonMobil, General Motors, and Toyota insist that they have made great progress. Whether and when those advances can make their way out of the laboratory is unclear, but in the unlikely event that gasoline reformation does take off, methanol will lose out. Gasoline is ubiquitous and familiar, and the world is already set up to deal with it on a massive scale.

Even if the technical complexities are mastered, the recent arrival of highly efficient hybrid cars—such as the Toyota Prius and the Honda Insight, which combine gasoline engines with electric motors—could still cause gasoline reformation to fail in the market. Although these cars are not exactly Porsches in terms of their horsepower, they were met in some quarters with unexpected enthusiasm: movie stars were spotted driving them in Los Angeles, devotees formed Internet chat rooms to celebrate their virtues, and huge waiting lists formed in dealerships across America. American car firms, which had initially pooh-poohed the Japanese efforts, promised to accelerate plans to develop their own versions. Hybrids, it appears, are a force to be reckoned with.

Robert Williams of Princeton University argues that as long as the main competition for the gasoline fuel-cell car was the conventional internal combustion engine car, the economic case for that type of fuel-cell car made some sense. The higher initial cost (an extra $5,000, including the gasoline reformer) could be more than offset by the likely doubling of the fuel efficiency plus the environmental benefits. However, the Japanese hybrids achieve levels of fuel efficiency and emissions comparable to those of gasoline fuel-cell cars today—but at a much lower cost. It is quite possible that gasoline fuel-cell cars could lose out to hybrids and thus fail to capture the market share needed to succeed. Such financial calculations, when combined with the engineering advantages of avoiding reformers, explain why the greenest option—direct-hydrogen fuel cells—is now the leading contender.

Seth Dunn of the Worldwatch Institute, an environmental think tank, argues that "the direct use of hydrogen is in fact the quickest and least costly route—for the consumer and the environment—toward a hydrogen infrastructure . . . There are no major technical obstacles . . . As one researcher has put it, 'If we really decided that we wanted a clean hydrogen economy, we could have it by 2010.' " Amory Lovins goes further. He insists that "the next revolution integrates efficiency with new supply technologies to provide mobility, comfort, and ultra-reliable power. It shifts the main energy delivery method from hydrocarbons to hydrogen. Today over two-thirds of the fossil-fuel atoms we burn are hydrogen, not carbon. The next step eliminates both the burning and the carbon."

That is the sort of prediction that always provokes loud guffaws from defenders of the status quo.

The Great Hydrogen Hoax?

The first reaction many people have when you mention any device involving hydrogen is "Will it blow up?" After all, using hydrogen as a fuel is an unfamiliar concept for most, while the hydrogen bomb is an all too familiar one. And there are those who have not forgotten the tale of the *Hindenburg*, the German airship filled with hydrogen that met with tragedy back in 1937. Some scientists and engineers are also skeptical of all the hydrogen hoopla, though for more technical reasons. Enoch Durbin, a professor of mechanical engineering at Princeton University, makes the following arguments:

> Automobile manufacturers are looking to develop fuel cell powered vehicles as a way of reducing our reliance on Middle East crude oil. Much of the popular press and many of our legislative leaders are enthusiastically embracing this new idea in energy, which they call the "Hydrogen-Fuel Cell Economy."
>
> Unfortunately, many are doing so without any real understanding of the technology they are embracing.
>
> *Hydrogen is not a source of energy!*
>
> Although it is the most abundant element in the universe, very little of it is available in a form we can use as a source of energy. To make hydrogen available for use as an energy source, we have to separate it from the elements to which it is bound, such as water (H_2O) or fossil fuels. This is somewhat like converting ashes of coal back into coal . . .
>
> *Why would anyone want to do this?*
>
> It is quite difficult to store hydrogen. The hydrogen molecule is so small that it tends to leak easily from any container in which it is stored. Even sophisticated engi-

> neers at NASA have trouble containing the hydrogen used in space propulsion systems . . .
>
> Once hydrogen is made it must be shipped to where it is needed. Distributing hydrogen is another problem. Typically gas or liquid energy is distributed by use of a pipeline. We do not have the pipelines to distribute hydrogen and they are extremely difficult to create. Electrical energy is transmitted over wires, a much easier distribution system.
>
> Why then would anyone go to all the trouble of making, storing and shipping hydrogen?

Why indeed! The honest answer is that unless hydrogen energy can overcome the three big hurdles raised by the skeptics—of safety, storage, and supply—it will not take off. Durbin would then be proved correct in concluding that "the use of hydrogen and fuel cells for general production of electricity is a bad idea approaching the ridiculous."

Despite the jibes from hydrogen skeptics that fuel-cell cars are "rolling H-bombs," the easiest of the three objections to tackle is safety. It is true that hydrogen is inflammable. But that risk has to be put into perspective. Unlike hydrogen, for example, methanol is corrosive and extremely toxic, while gasoline is both a carcinogen and easily ignited. A study done in 1997 by Ford argued that hydrogen-powered cars, if properly engineered, could potentially be safer than those using gasoline or propane. In fact, as a top auto executive pointed out to a congressional panel, if gasoline had to win approval by environmental authorities as a new fuel today, it could well be rejected as too dangerous. A related factor is that hydrogen is a gas at room temperature and disperses rapidly, unlike methanol and gasoline.

As for the theory that the *Hindenburg* exploded because it was filled with hydrogen, it is almost certainly wrong. That same ship, as well as a sister dirigible, had by 1937 made many transatlantic crossings without mishap. More to the point, the ship never exploded; it caught fire. Some terrific detective work done by Addison Bain, the former head of the hydrogen program at the Kennedy Space Center, revealed that the cause of the tragedy was very likely to have been static electricity sparking some of the highly flammable compounds with which the Germans had painted the ship—compounds that are today used to propel modern rockets. The lesson, according to Bain, is not that hydrogen vehicles are destined to blow up; rather, he concluded dryly, it is that you "don't paint your airship with rocket fuel."

A tougher challenge for hydrogen is storage. Because hydrogen has the smallest atomic structure of all elements, its atoms can wiggle through the crystal lattice of the material used to contain it. The leakage from a pressurized hydrogen tank could be significant. Being so small, hydrogen is also exceptionally light. In a typical gaseous storage system, it has only a tenth of the volumetric energy density of gasoline. That lightness can be a good thing in itself, but it does means that hydrogen takes up a lot of space.

The obvious solution is to compress the hydrogen. Impco, an American firm, devised an ingenious all-composite tank that could hold enough hydrogen at a pressure of 5,000 pounds per square inch (psi) to allow a small fuel-cell car to travel three hundred miles. The tank met stringent safety standards and was expected to cost less than $1,000 once commercialized. Holding more than 40 gallons of hydrogen, however, it was still far bulkier than the average gasoline tank. But the firm and its rivals were also developing tanks capable of storing hydrogen at 10,000 psi, which are more compact. Even so, the Holy Grail of hydrogen storage is in some solid form, which would offer advantages in safety as well as convenience.

An Unbearable Lightness

"There may be chimneys out there somewhere producing this stuff right now!" Nelly Rodriguez exclaimed enthusiastically as she bustled around her laboratory at Boston's Northeastern University on a crisp winter afternoon. The object of her affection was a crumbly, black bit of soot that looked as though it had indeed come straight out of a chimney. She was excited about the possibly revolutionary, though as yet unproved, potential of carbon "nanotubes"—an elemental form of carbon that experiments suggest could store and release astounding quantities of hydrogen.

It has long been known that solid materials called metal hydrides are capable of storing small amounts (about 1 to 2 percent of their own weight) of hydrogen at room temperature. Some metal hydrides are capable of storing more hydrogen (5 to 7 percent of their own weight) but do so only at impractical temperatures of 250°C or higher. Carbon nanotubes and "nanofibers," which were discovered only a few years ago, seem to be able to absorb hydrogen just as well, even at room temperature. This opens up the prospect of using these materials, where each grain is a tiny carbon "sponge" like the one Rodriguez got excited about, to store hydrogen and release it on demand. America's Department of Energy has calculated that carbon materials will need to store at least 6.5 percent of their own weight of hydrogen—retrievably and at ambient temperatures—in order to make fuel-cell cars practical (defined as having a range of three hundred miles or so between refueling stops).

For a brief moment in the late 1990s it appeared that the hydrogen breakthrough had happened. Nelly Rodriguez and Terry Baker, an Englishman who was her husband and academic collaborator at Northeastern University at the time, achieved an astounding 65 percent storage rate with one batch of nano-materials. The entire hydrogen world stood at attention. Everyone knew what was at stake: if those results could be reproduced, then the storage problem would be cracked—and the first to market would mint a fortune. Such materials could be used to make hydrogen cartridges that would slot into fuel-cell cars, making refueling as simple as swapping an empty cartridge for a full one. The snag is that nobody has been able to reproduce anything close to those numbers.

When I visited their cluttered laboratory, Rodriguez and Baker were very pleasant but also extremely secretive. When I asked why their results could not stand up to peer review, they would say only that the big commercial clients that they had talked to (under strict vows of confidentiality, naturally) were plenty satisfied. Most scientists remained skeptical of the dynamic duo, even though many remained hopeful about carbon nanotubes generally. These materials are so complex that it is difficult to predict their properties theoretically. Perhaps the explanation for their behavior lies in entirely new and unanticipated kinds of hydrogen-carbon interaction. If so, then Baker and Rodriguez, now widely considered either frauds or flukes, may come to be seen as the father and mother of the hydrogen age.

If nanotube storage is perfected, the entire hydrogen economy would leap forward by decades. Until then, however, the more immediate solution remains metal hydrides, which store and release hydrogen in the way that the batteries in some of today's mobile phones and laptop computers do. The firm that pioneered the rechargeable nickel-metal hydride battery is ECD Ovonics of Michigan. Stanford Ovshinsky, its colorful and opinionated boss, thinks he can repeat the trick for fuel cells; he vows to get a practical metal hydride tank on the road by 2005 or so. That is highly ambitious, given the lingering problems with hydrides, but he has managed to persuade some business heavyweights that it might be an achievable target. The oilmen at Chevron Texaco, for example, bought 20 percent of ECD and entered into joint ventures in the areas of fuel cells, hydrogen storage, and metal hydrides. Robert Stempel, the former boss of General Motors, was so excited by Ovshinsky's vision that he came out of a cozy retirement to become chairman of ECD Ovonics.

Rivals are working on all sorts of other approaches to storage, ranging from gaseous to chemical to solid state, that promise a flurry of innovation in coming years. One firm has even achieved modest success using the inexpensive chemicals found in that old laundry detergent, Borax. Sound crazy? Not to DaimlerChrysler, which unveiled Natrium, a prototype fuel-cell vehicle, using precisely that sudsy technology for hydrogen storage. This minivan got its juice from a 20 percent solution of sodium borohydride (a hydrogenated version of the stuff in the Borax box) in water. Stepping on the gas pedal pumped this chemical cocktail past a catalyst, which forced the hydrogen out of the solution and into the fuel-cell engine. The biggest handicap of this clever approach was that it produced loads of slurry, which company officials figured could be emptied whenever the driver refueled her car.

Nobody really knows which particular technology or company will hit the hydrogen jackpot. However, given all the frenzied activity, it seems pretty plain that the storage problem need not be insurmountable.

Pond Scum to Petro-Hydrogen

As Enoch Durbin pointed out, hydrogen fuel is not an energy source. In itself, hydrogen is just an energy "carrier." In other words, it is more akin to electricity—which must be produced by somebody, somewhere, using a primary energy source like coal or gas or wind—than it is to petroleum. Hydrogen is the most abundant element in the universe, but it rarely exists in its free state on earth, being found normally in combination with oxygen (as water) or carbon (as methane and other hydrocarbons). As a result, it always takes energy to free it for use, whichever way it is produced.

Making electricity from primary energy sources is obviously still worth the effort. Similarly, using an energy carrier like hydrogen would save us from carrying around giant sacks of coal, or windmills attached to the roofs of our cars. Gasoline also serves this purpose today—but it is a dirty fuel that cannot be used without harming the environment and human health. Hydrogen, in contrast, can power a car while producing no harmful emissions. That is why visionaries see a clean energy future in which hydrogen and electricity will co-exist, each one used in the tasks for which it is best suited.

There are several competing strategies for getting hydrogen-based energy off the ground. One approach is electrolysis, which zaps hydrogen free from water by using electricity. This process is energy intensive, so large-scale electrolysis is likely to take off first in places with cheap, clean sources of energy, such as geothermal springs or hydroelectricity. However, the electrolysis approach does benefit from the fact that its two prerequisites, electricity and water, are fairly well distributed around the world.

In the long term, proponents argue, the world will get its hydrogen directly from renewable energy. Many research projects are under way around the world that aim to use the energy derived from the wind or the sun to produce hydrogen directly (by electrolysis of water). According to one study, a small fraction of the land area of a sunny place like Mississippi or Spain set aside to produce solar hydrogen would be enough to power a fleet of 200 million fuel-cell cars. At what price? The fans of such technology would prefer not to say, at least until costs come down from the stratosphere.

However, field tests are helping their cause. In California, the SunLine Transit Agency (which serves Palm Springs and its desert environs) boasts of being a world leader in converting its fleet to hydrogen: "Working with such manufacturers as Cummins Engine Company, Detroit Diesel, Engelhard Corporation and John Deere, SunLine has become a continuous beta test site for clean air equipment innovations." SunLine already has a facility that uses solar energy to produce hydrogen, and it is planning another that will harness wind energy. In 2001 Honda also opened a solar-powered hydrogen production and refueling station at its research labs near Los Angeles.

Once produced, hydrogen could also be used as a form of energy storage. Power generated whenever the wind blows can be stored as hydrogen in a tank and sold into the power grid when needed or only when peak power prices are offered. That would revolutionize the way electricity trading is done. Electricity is one of the very few commodities that cannot be stored easily, and it is almost always used or transmitted as soon as it is generated. The European Commission's grand "Hydrogen Road-map," unveiled in mid-2003, saw hydrogen storage as the essential stepping-stone for its very ambitious renewable-energy plans.

Some academics are working on an even more radical concept, called biolysis, that involves manipulating the metabolism of algae and other life-forms to release large quantities of hydrogen when they decay. Others are tinkering with photolysis, which harnesses sunlight directly to split water without the need for electrolysis. Fascinating as such approaches are, they are ridiculously inefficient ways to produce hydrogen. For the moment, then, the world's biggest source of hydrogen is likely to be commonplace fuels like natural gas and coal.

The idea of making hydrogen by stripping it out of such hydrocarbons is unpopular with those environmentalists who oppose everything involving fossil fuels, but it is hardly a foreign concept. The world already consumes large quantities of hydrogen produced today by reforming natural gas at centralized plants. The hydrogen is used to make ammonia fertilizers and to "lighten" heavy grades of crude oil. Though few people probably realize it, hydrogen is essential to the production of the margarine they spread on their toast at breakfast.

The great attraction of the above approach is that the earth's vast reserves of filthy coal, which are now thought of as environmental enemy number one, could be tapped to produce clean hydrogen fuel. Using coal in such a way would do much to propel the world toward a cleaner, low-carbon future. Consuming hydrogen from coal in fuel-cell cars and micropower plants would produce no noxious emissions at the point of use. The drawback is that a lot of carbon dioxide (CO_2) would inevitably be produced during the reformation process (which would probably take place at big plants located outside cities). Scientists are coming up with a very promising array of technologies—collectively known as carbon sequestration—to capture that CO_2 and store it safely somewhere for a very long time so that it does not enter the atmosphere.

Some greens raise serious objections to sequestration. An article that appeared in *The Ecologist* by Greg Muttitt and Ben Diss of the NGO Platform put it in almost spiritual terms. The authors lambasted geological carbon sequestration (the underground storage of CO_2 in such places as oil reservoirs) as typical of the macho, technophilic men at the top of the oil industry: "They are part of a very masculine culture that is obsessed with technological toys, that believes that bigger is better, and that cares little for anyone's interests but its own. In this context, it's hardly surprising that it came up with the idea of drilling 1,000 meters into the ground to pump down carbon: what could be more masculine than that?" Muttitt and Diss conclude with this impassioned appeal to kill sequestration now: "But theirs is a false god . . . geological carbon engineering leads us deep into the unknown, and carries enormous risks. It is a dangerous diversion and needs to be treated with great caution."

Robert Socolow, an expert at Princeton, disagrees. He argues that "the politics of fossil-carbon sequestration are unlike the politics of carbon management strategies designed to bring the fossil fuel era to a rapid close. The fossil fuel industries are willing participants, and they are showing leadership. So are many countries and portions of countries rich in fossil fuel resources. The result should be new coalitions supportive of policies intended to mitigate climate change." Sequestration should be allowed to compete with other approaches designed to tackle climate change, he says. "As a carbon management strategy, fossil-carbon sequestration is in competition with the substitution of renewable energy for fossil fuels and with the substitution of nuclear energy (fission and fusion) for fossil fuels. At this time, one can only guess how the three strategies will compete. My guess is that for the next hundred years all three will co-exist, each of them contributing substantially to carbon management."

Sequestration could well act as a stepping-stone to a renewables-based hydrogen economy. An unlikely political coalition seems to be emerging in favor of such a multipronged approach, much as Socolow predicted. Eight big energy companies, including some oil giants, have banded together in the CO_2 Capture Project to promote research in this area. The Natural Resources Defense Council, an American green group strongly opposed to the use of biological carbon sequestration (i.e., the use of agricultural lands and forests as carbon sinks), now says it is willing to keep an open mind about geological carbon sequestration. The most important endorsement, though, goes as follows: "We all believe technology offers great promise to significantly reduce [greenhouse-gas] emissions—especially carbon capture, storage and sequestration technologies." Those words come from none other than George W. Bush, the man who said no to the Kyoto treaty on climate change. When such a stout defender of fossil-fuel interests began supporting sequestration, the prospects for a hydrogen transition brightened considerably.

Care for an Espresso—or Would You Prefer to Suck on the Tailpipe?

Boosters of fuel-cell cars would really like you to watch the mayor of Chicago swallow tailpipe exhaust. It's not that they have anything against him; in fact, Richard Daley is one of the industry's favorite political leaders. He welcomed several hydrogen-powered buses onto Chicago's streets as part of a pilot program, and to prove that these are some really, really clean buses, he even drank a glassful of steamy exhaust coming straight from one of the tailpipes. Using pure hydrogen with no need for reformers, these buses delight those like Amory Lovins who argue for the direct-to-hydrogen approach.

To reduce the cost of manufacturing fuel cells, and to win public acceptance, a number of central and local governments will have to follow Mayor Daley's lead in encouraging a shift to hydrogen for fleet vehicles such as city buses, delivery trucks, and so on. The size of these vehicles will keep the bulkiness of tanks for compressed hydrogen gas from being a significant penalty. Fleets of commercial vehicles have the added advantage of refueling at central depots, so setting up the infrastructure for refueling them is less of a problem. Hydrogen buses have found their way onto the streets of Vancouver, Stuttgart, and parts of California as well as Chicago. The World Bank is now funding the rollout of fuel-cell buses in megacities from Mexico to China. In 2003, both Federal Express and UPS put fuel cells into some of their delivery trucks.

The oft-cited estimates of $100 billion or more for phasing in hydrogen infrastructure are outlandish. They assume that today's gasoline infrastructure must be duplicated from day one, which is not the case. Experience with the introduction of diesel in America and unleaded gasoline in Germany shows that even if only 15 percent of gas stations offer it, a new fuel can become widely accepted.

Tapping into the existing natural-gas grid to reform hydrogen locally, rather than shipping it via pipeline, looks like the smartest way to reduce infrastructure costs. Firms such as United Technologies, a multinational with decades of fuel-cell experience, and even oil giants like Chevron are now developing small reformers to do precisely that. This type of reformer can be placed at gasoline stations, supermarkets, or even office blocks. Stuart Energy, a firm based in Toronto, is building tiny electrolyzers for a car that can produce hydrogen by splitting water using off-peak electricity straight from the wall socket. Electrolysis is especially suited to the early years of hydrogen-powered cars because the technology involved can start small and scale up as demand grows.

Sandy Thomas of the pioneering firm H2Gen Innovations makes a compelling case about the viability of a hydrogen rollout:

> A hydrogen infrastructure based on reforming natural gas at the fueling station for use in fuel cell vehicles would be less expensive than the existing crude oil-to-gasoline fueling system, with the potential to reduce global cumulative motor vehicle fueling infrastructure investment costs by US$ 840 billion to US$ 1.1 trillion over the next 40 years . . . Hydrogen made from natural gas that is used in fuel

> cells can substantially improve our homeland security and reduce local pollution and greenhouse gas emissions while providing a pathway to an eventual sustainable energy future based on renewable hydrogen. However, private industry has little incentive to expeditiously implement a hydrogen and fuel cell energy system when the primary advantages—improved security and a cleaner environment—accrue to society as a whole.

Despite the many attractions of a hydrogen economy, Thomas still thinks that direct government intervention may be necessary to make it happen.

That view is shared by the team of hydrogen experts at Princeton University. Robert Williams, Joan Ogden, and several of their colleagues crunched the numbers on fuel-cell economics and came up with a dramatic conclusion: after the initial introduction, direct hydrogen fuel-cell cars will offer significantly lower costs and greater benefits to their users, as well as to society as a whole, than rival fuel-cell options. Unfortunately, the initial hurdle is so high that market forces alone may not spur the necessary investments. The Princeton team thinks the direct-to-hydrogen route will fail unless governments embrace zero-emission mandates like California's initiative.

Of course, an alternative to government mandates is the introduction of higher gasoline taxes in the United States, on the justification that the harm done to the environment and human health by petroleum is not accounted for in the current price (never mind the cost of overreliance on the fickle OPEC cartel). Leveling the playing field in that market-friendly way would also boost hydrogen's prospects. If fuel cells really take off in cars, the Princeton group envisions extremely low vehicle emissions within twenty years—and consumers would pay no more than they do today for transportation.

Even Lance Ealey and Glenn Mercer, who penned that skeptical piece in *The McKinsey Quarterly*, agreed with Robert Williams on this point. They acknowledge that "emissions regulation is the Achilles' Heel" of the conventional car engine, but they go on to make this gloomy forecast: "Only regulation could facilitate a quicker transition to the fuel cell. Indeed, environmental concerns about greenhouse gas emissions such as carbon dioxide and the geopolitical desire for energy independence may accelerate the demise of the internal-combustion engine, for governments could enact policy reforms to favor the development of alternative technologies and their adoption by the consumer. But until the consumer is ready to embrace them, most governments are unlikely to accept the political risk of radical reform." No sooner had those words been published than just such an "unlikely" thing happened: California governor Gray Davis signed into a law a controversial and far-reaching act that—for the first time ever—would regulate tailpipe emissions of carbon dioxide. If that law stood up in court, said legal experts, other states could be expected to follow suit.

Ealey and Mercer got it half right: whatever governments do, fuel-cell cars will never make it big unless they offer benefits that ordinary consumers actually want. The clearest advantage of fuel cells over the internal combustion engine is the potential for very low or zero emissions. But that may not be important to consumers. After all, more than half of new vehicles sold in America (and a rising share in Europe) are gas-guzzling pickup trucks and SUVs. A more immediate benefit is likely to be the fuel cell's superior efficiency. Today's internal combustion engines are notoriously inefficient, converting only about 15 percent of the heat content of gasoline into useful energy. Even in their current primitive state, fuel cells manage at least twice that. As fuel-cell technology matures, a rising level of efficiency will mean falling operating costs.

Consumers may find other features of fuel-cell cars attractive, such as a nearly silent ride. How much sweeter will your Beethoven sonatas or gangsta rap tracks sound when not competing with the grunts and groans of your transmission and gasoline-burning engine? Another nifty advance is that fuel cells offer a clean, "engine off" energy source for power-hungry electronics. That means that advanced electronics—from computerized tush-warmers to GPS satellite-tracking systems to space-age audio and video equipment—can be powered from an electrical power source far more robust and reliable than the one found in today's cars.

The Detroit Auto Show held in early 2002 showed that the switch to fuel cells could inspire a top-to-bottom redesign of the entire concept of the automobile. After dismissing efforts like those of Amory Lovins for years, General Motors did a sharp U-turn and took a page right out of the Hypercar handbook. It unveiled a concept car called AUTOnomy that, to the ordinary person, looked like a giant skateboard the width and length of a normal car but less than a foot thick. In fact, inside that low-riding package were the necessary fuel cells, fuel-storage system, suspension, drive motors, and electronic controls needed to power the car of the future. By bundling all of the car's innards in that sleek skateboard chassis, the firm was able to replace many conventional mechanical interfaces with electronic controls that are cheaper, more efficient, and more exciting to use.

GM's designers relied on several insights that promise to revolutionize your relationship with your car. First of all, there's no steering wheel: because fuel-cell technology is perfectly compatible with advanced "fly-by-wire" steering systems, you will drive these cars of the future using controls similar to those used for video games or jet fighters. Second, because advanced electronics will replace mechanical systems like brakes, throttle, and steering that are prone to wear and tear, your car will be extremely cheap to maintain. Finally, because the AUTOnomy chassis can easily be fitted with various types of car bodies, it means you can change the shape, color, and feel of your car without having to buy an entirely new one. Who knows? In the fuel-cell future, you may be changing car bodies seasonally the way that fashion victims today change wardrobes.

But wait, there's more. As Amory Lovins has long argued, the fuel-cell car could even be a source of revenue for homeowners.

Plugging it into your home's electricity supply and transmitting the power generated by the car back to the grid while it sits in your garage at home or in the office could earn you a profit on the energy market. He sketches out his direct-to-hydrogen, no-subsidy-required strategy for fuel cells thus:

> The key is to integrate deployment in cars and buildings so that each makes the other happen faster. For example, Hypercars can be leased initially to people who work in or near buildings where fuel cells will by then have been installed for power generation and space-conditioning. The cars can be designed to serve as plug-in power plants when parked (about 96% of the time)—buying surplus hydrogen from the building, where it's typically made from natural gas, and selling fuel-cell electricity back to the grid at the time and place where it's most valuable. This can repay much or most of the cost of owning the cars. It could ultimately provide 6–12 times as much generating capacity as all electricity suppliers now own, displacing the world's coal and nuclear plants many times over.

Lovins jokes that the key consumer benefit could even be the option of having fine espresso brewed on the dashboard, using the fuel cell's exhaust of steaming water.

All this explains why going direct to hydrogen is the best approach. However, it is still possible that firms betting on such a technically superior option may get trumped in the marketplace by rivals peddling an inferior but more accessible technology—say a gasoline-engine hybrid. That would be a replay of the VCR wars of the 1980s, when Sony's technically superior BetaMax format was defeated by the shoddier but ubiquitous VHS format. Don Huberts, the former boss of Shell's hydrogen division, handicapped the race this way: "Everyone is placing bets on several horses. By no means is it clear today which the winner will be."

Fuel Cells Meet Big Business

The moment when an experimental technology becomes a commercial one is hard to define, but the interest of oil companies, carmakers, and power-engineering firms—almost all the industries that have a stake in the business—is a sign that fuel cells are crossing the line. Cozy incumbents in most industries typically resist new technologies that threaten to make their existing capital stock worthless, but even Shell now has a hydrogen division. The very notion of an oil giant investing serious money and credibility in such a technology would have been laughable just a few years ago. Yet today, Sir Philip Watts, the chairman of Shell, does not even blush when he forecasts a "decarbonized energy world" based on "hydrogen energy and fuel cells." If the big boys of the energy business think that fuel cells are coming, they probably are.

Why did they turn from obstructionists to enthusiasts? Whatever their clever advertising campaigns say, the reason is not because Big Oil suddenly decided to worry about the environment. The real reason is that recent technical advances have been so promising that the incumbents simply could not ignore hydrogen any longer. One oil boss explains: "Fuel cells have produced more technological breakthroughs in five years than battery research has in the past thirty." The advances are so great that market incentives, not mere regulation, are now motivating them. As Texaco's Graham Batcheler put it (before his firm was swallowed up by Chevron): "We came around late to fuel cells, but we now recognize that the oil and gas business is going to change . . . whatever fuel emerges eventually as the choice for fuel cells, we want our consumers to fill up at a Texaco station." In other words, even the traditional purveyors of fossil fuels, realists to a fault, now believe in fuel cells. For a technology that has depended on visionaries for 150 years, that is quite a breakthrough.

A Northern Star Leads the Way

Pay a visit to Firoz Rasul in Burnaby, a small town outside Vancouver, Canada. Rasul made his way from Kenya to Europe, where he studied and worked. He then earned an M.B.A. at Canada's McGill University and rose through the ranks at various firms before arriving at Ballard Power Systems in 1988. In his own quiet way, he has transformed Ballard beyond recognition.

Ballard got its start doing contract research, including top-secret stuff for the Canadian military. In the early 1990s Ballard decided to start nurturing proton-exchange membrane (PEM) technology into a commercially viable product for the civilian market. The original research patents in this area were held by GE, but they had expired by the mid-1980s. Ballard's first brilliant move was to swoop in and register its own patents. Then the firm dedicated top-tier scientific talent and money to commercializing the technology. Ballard now holds hundreds of crucial patents on PEM and related technologies. Under Rasul's leadership, it has gone from being an obscure contract-research laboratory to a technology powerhouse in automotive fuel cells.

But Rasul has bigger ambitions: he wants to turn Ballard into the dominant force of the emerging clean energy industry. *Red Herring* magazine ran a flattering piece on Ballard entitled "The Next Intel?" In it, Rasul explained his strategy for world conquest:

> "We have studied how Intel has written the rules for the microprocessor industry," Mr. Rasul says. "They used their position as the leaders of a hip technology to create the strategic relationships they needed to establish a key position." Ballard has used its position as the developer of the PEM fuel cell to build joint ventures with Ford Motor and DaimlerChrysler, sold the technology to key competitors like General Motors and Toyota, and kept its focus on intellectual property by registering over 350 patents. "The

> point is that with any enabling technology such as the fuel cell or the microchip, you've got to put the technology in the hands of the people who can develop products, even if they're your competitors," says Mr. Rasul.

So impressed were big auto firms by Ballard's PEM technology that two of them, Ford and DaimlerChrysler, even bought equity stakes in the company and arranged for special joint ventures with Ballard.

This field is young and fluid, and it would be foolish to make any concrete predictions about Ballard; the firm faces stiff global competition and could yet go bust. Indeed, investors drove down Ballard's share price by 20 percent in a panic immediately after *Barron's* ran an article in July 2002 suggesting that the firm was burning through an awful lot of cash, considering that its products are far from commercialization. Even so, if any fuel-cell firm has a chance of success, it is probably Ballard. Remember that zippy NECAR 5 that went from California to Washington? You bet it was powered by Ballard fuel cells. The firm has even adopted a slogan inspired by the ubiquitous "Intel Inside" tag line: "Powered by Ballard."

Nearly two years before the widely circulated *Red Herring* profile appeared, Iain Carson, the industry editor of *The Economist*, penned a story called "Intel on Wheels" that nailed Ballard perfectly. However, even he could not crack one part of the story. The article starts like this:

> Visitors to Ballard, a chemical-to-cars firm based in Burnaby, outside Vancouver, are welcome to walk around the laboratories and workshops where small-scale manufacturing of fuel cells is under way. They are free to talk to the chemists who have found cheaper ways to make the components of a fuel cell. Here is the polymer membrane that used to cost $750 a square foot; and here also the graphite frame that cost $100 a few years ago. Ballard can now make both for only $5 each. But along one side of the laboratory

is a windowless, cream-painted plywood wall; and what happens behind it is strictly off-limits.

Behind that wall lies Plant 1, the top-secret bunker from which Ballard intends to conquer the world. Two years after Carson's visit, I became the first journalist ever to penetrate this Fort Knox of the fuel-cell world.

With a cup of strong coffee in hand, I made my way to Ballard's headquarters at an ungodly hour in order to sneak a peek into its top-secret manufacturing facilities. To be quite honest, Plant 1 was unremarkable. The facility's expansive shop floor was filled mostly with standard manufacturing equipment, the arrangement was typical of assembly lines everywhere, and the early morning shift of factory workers filed in bleary-eyed for another day of humdrum work. In other words, the place warmed up in the morning just like millions of mass-production facilities all over the world.

It was only when I found the firm's research building a few hundred yards away that it became clear exactly how extraordinary Plant 1 really is. The laboratories are the site of dramatic breakthroughs that have made fuel cells so tantalizing. In just the last decade or so, Ballard's scientists have managed to shrink a stack of PEM fuel cells powerful enough to run a small car from the size of a giant fridge to the size of a microwave. They have reduced the amount of pricey platinum required by several orders of magnitude, and they have otherwise redesigned the stacks for ease of manufacture and assembly.

The result: Ballard now hopes to produce many thousands of units per year from Plant 1. Only when you see the graying hydrogen gurus tinkering away patiently amid the clutter of laboratory equipment next door does the picture become clear. Taking a page from Ford's visionary embrace of the assembly line, the real advance in Plant 1 is the slow but sure shift toward volume manufacturing of fuel cells. And for that, credit must go to Rasul. One of the first things he did when he arrived at Ballard was to light a fire under those brilliant physicists and electrochemists: "If you guys are looking for a Nobel Prize, you are in the wrong place. If you are looking to make a lot of money, you are in the right place." In early 2003 Rasul retired to the ambassadorial position of executive chairman and handed over day-to-day management of the firm to Dennis Campbell, an outsider picked specifically because of his experience in helping technology companies with what he called "mass commercialization." The shift taking place at Ballard and its rivals is the best reason to think that the time for fuel cells has come.

Still not convinced? Consider these words from a new hydrogen convert: This technology "will fundamentally alter the American way of life in a positive way!" So said George Bush as he unveiled his $1.2 billion hydrogen strategy in early 2003 to a roomful of executives from the energy and automotive industries. His announcement is especially striking when you consider the play on the defiant words used by his father at the Rio Earth Summit a decade earlier—the American way of life, the elder Bush had insisted, was not up for negotiation. Of course, a billion dollars will not do much to persuade industries that have hundreds of billions of dollars in sunk investments in fossil-fuel technology (and which now plan to spend many billions more on oil exploration, as the next chapter explains). Bolder moves—like an end to fossil-fuel subsidies and a carbon tax—are surely needed.

Even so, hydrogen energy got an undeniable boost when the Texan oilman in the White House gave it his unqualified endorsement. All that was missing from Bush's rallying cry was the energetic banging of some tom-toms.

9

Rocket Science Saves the Oil Industry

"FIVE TO TEN YEARS AGO, if you came to me with the idea of a directional well that stretched 25,000 feet . . . well, I'd have said you're dreaming." Brian Kuehne paused for a moment and looked around the room at his fellow oil drillers. These men were used to the rough-and-tumble of fieldwork, and being cooped up in that stuffy conference room at Royal Dutch/Shell's skyscraper on the outskirts of the French Quarter of New Orleans made them uncomfortable. Still, they nodded vigorously in agreement. "Today—" began Kuehne, pausing for dramatic effect. "Today, we're just kicking them down one after the other!"

Kuehne and his colleagues from the drilling side of the business had gathered at the firm's regional office, which manages operations in the eastern Gulf of Mexico, to explain the impact of recent technological advances in oil exploration and production (E&P). Through fancy charts and slides, potted histories, and personal anecdotes, these old hands claimed that their business was being radically transformed.

I decided to visit Shell's Ursa platform out in the deep waters of the Gulf of Mexico to take a look for myself. Sure enough, those drillers turned out to be wrong: they were far too modest. In fact, the technological changes now sweeping through the oil business promise to transform it almost beyond recognition.

The $1.5 billion Ursa platform is one of the most sophisticated in the world. Its remarkable "tension leg" design allows it to sit safely atop 3,800 feet of treacherous, hurricane-infested water—a depth thought unconquerable just a few years ago. The floating city of steel is also so heavily instrumented that its control room is far more evocative of the U.S.S. *Enterprise* on *Star Trek* than the grimy gushers of yesteryear.

Ursa pumps so much oil that it more than paid for itself in its first three years of operation. And there is plenty more gravy yet to come. The day I stood gawking atop one of mankind's great feats of engineering, knees trembling with every slight quake and quiver of the floating platform, the crew achieved something really breathtaking. With the help of advanced robotics and seismic know-how, it drilled an elaborate multidirectional well that twisted and turned its way around obstacles beneath the ocean floor to hit a giant pocket of oil 28,000 feet away.

"We've drilled wells in the same way for a hundred years," explained Raoul Restucci, boss of Shell's E&P Company, "but in just the last few years we've seen dramatic changes in technology that are greatly reducing the cost of accessing a molecule of oil." Restucci's firm is responsible for extracting two-thirds of all the oil ever produced from the deep waters of the Gulf of Mexico, and it is the prime force behind Ursa. The secret to Shell's success in such risky realms, he reckons, is technology—and the business thinking that ensures that the firm profits from that technology.

Shell's technology leadership is not unchallenged. ExxonMobil, the only other oil major investing seriously in proprietary E&P technology, is also producing successfully at ultra-deep water platforms of Ursa's sophistication. At the secretive firm's discreet research laboratories in Houston, located just a stone's throw away from Shell's own much trumpeted Bellaire research institute, a team of top Exxon researchers have come up with seismic imaging techniques that allow highly sophisticated visualization of reservoir dynamics in minutes—rather than the months the same job required a few years ago. The firm's technology gurus put on an impressive three-dimensional show that modeled an actual reservoir in real time. Tucked away behind the snazzy semicircular screen was nearly $80 million worth of supercomputers and other high-end hardware known affectionately as the company's "analytic brain."

Asked whether claims about recent technological change are exaggerated, one Exxon engineer's response came swiftly. The oil industry, he insisted, "is on the cusp of dramatic breakthroughs in exploration and production!"

While his excitement is understandable, the pace of innovation in petroleum is not really surprising when you consider the economic prize. The global economy is utterly dependent on this filthy, geographically concentrated and geopolitically risky hydrocarbon to run its planes, buses, and cars. What's more, this industry receives enormous subsidies and disguised incentives (ranging from the undertaxation of gasoline in many countries to the West's military presence in the Middle East, to free dredging of canals by America's Army Corps of Engineers for the safe passage of oil tankers). All of these perversions of the market reinforce the petroleum economy and reduce the financial risk that the oil sector must absorb. This tilted playing field as much as anything explains why many oil giants are ready to throw tens of billions of dollars into far-flung oil fields or speculative E&P technologies sooner than they will consider investing minuscule sums into clean energy.

That sorry state of affairs discourages many greens, and rightly so. Still, they should not despair. If they look back in history, they will find that petroleum itself was once such an unproved energy upstart, fighting all sorts of incumbent technologies and favored fuels. And it was not subsidies or state intervention that helped it get off the ground: it was entrepreneurship, ingenuity, and a dash of inspiration.

Oil's Last Frontier

More than a century ago the industry's early technologists, in their quest for petroleum, adapted the Chinese innovation of drilling into the ground for salt. It was Edwin Drake, a curious character who passed himself off as the Colonel, who drilled America's first oil well in Titusville, Pennsylvania, in 1859. Local farmers, who had only ever scooped out small quantities of the stuff as it oozed out of rocks in the area, ran about screaming, "The Yankee has struck oil!"

His discovery set off the country's first oil boom, but its impact was limited. The internal combustion engine had yet to be commercialized, and the oil produced was used chiefly to displace whale oil for lighting. Also, Drake's first well hit oil at a depth of less than 70 feet, and produced only around 30 barrels a day. Though it was an impressive and unexpected achievement, it paled in comparison with the raging gusher that was to come four decades later at Spindletop.

A casual visitor to Beaumont, Texas, today could miss altogether the fact that this was the birthplace of the modern oil industry. The town's crumbly downtown area and generally sleepy character suggest that it has always been a backwater. Those who make the trip to the Gladys City museum—a re-creation of the original boomtown that is lovingly maintained by local history buffs—might get a taste of the revolutionary technology that Pattillo Higgins and Anthony Lucas brought to bear on an obscure hilltop outside town.

Higgins, a local man who liked to picnic with his Sunday school students on Spindletop Hill, had long been convinced that there was a vast oil deposit there. Though many ridiculed his ambitions, mocking him as the village "millionaire," he bought up as much land on the site as he could afford and began drilling with unsophisticated technology. His money began to run out before he struck any oil, and he advertised for a new investor. The only taker was another odd character, Anthony Lucas, who had been a captain in the Austrian navy. But he too found that drilling with conventional technology got him nothing but empty pockets. Eventually Lucas was himself forced to turn to yet another set of investors for help, confessing that "there have been three attempts to drill wells on that hill by practical well men, and no one has been successful in getting deeper than 400 feet." His savior replied, "Captain, I know a man who can drill that well to 1200 feet if it can be done at all. His name is Jim Hamill."

Hamill and his brothers were pioneers in using a revolutionary drilling technology: fancy drill bits that rotated their way through, rather than merely pounding the heck out of, the earth below. Months of frustrating work, including a number of setbacks, nearly killed the project. Then, on a chilly morning in early January 1901, they struck black gold.

The Spindletop Gusher, a popular history book sold by the Gladys City museum, describes the event this way: "As they lowered the sharpened bit into the 1020-foot hole, a strange hissing noise filled the air. Suddenly, a thick plume of mud shot up from the bowels of the earth, and with it came 4 tons of drilling pipe. The men ran for their lives, dodging the huge chunks of metal that rained down around their heads." Frustrated, and convinced that this was a catastrophe, the men waded back knee-deep in mud to clean up the mess. At that moment came another explosion: "With a huge, deafening roar, greenish black oil spurted from the ground. The geyser of oil climbed higher and higher, until it reached 100 feet above the top of the wooden derrick . . . [It was] the greatest oil well that the world had ever seen." Spindletop gushed for nine days and nine nights before it could be brought under control, spewing forth perhaps 100,000 barrels of oil each day.

That raging well transformed Gladys City (the humble shacks around Spindletop that Higgins had named after a little girl in his Sunday school class) into a boomtown worthy of the California gold rush. By spring of 1901, more than 130 oil wells were crammed onto that hill. The population of Beaumont exploded from 9,000 to more than 50,000, with more opportunists arriving every day. Land that Higgins had bought for just $6 an acre a few years previously was selling for as much as $1 million an acre. More than any other single event, the boom at Spindletop announced the arrival of the modern oil industry.

Technological innovation was bred out of sheer necessity back at Spindletop, and so again it is in today's industry. Only this time the industry is drilling not only deep under the earth but through the depths of the ocean too. Part of the explanation for the timing of today's technology revolution clearly lies in the enormous promise of deepwater exploration—the oil industry's last frontier. The development of such breakthrough technologies as advanced seismic imaging encouraged firms to venture into such inhospitable (and, not too long ago, unpromising) terrain.

The wheel has now come full circle: dramatic early success in places like Ursa is in turn driving technological innovation. That is adding perhaps a couple of decades' worth of life to the oil industry. What happens to the oil business after that, however, will depend less on technological wizardry than on difficult energy-policy choices now facing political leaders and ordinary folk around the world—but especially in America, the biggest importer and consumer of petroleum.

With all the continents save Antarctica poked and probed to death over the last century, experts are convinced that there are few giant "elephant" fields left to be discovered on land. Even the discoveries in countries surrounding the Caspian Sea, often talked up by journalists and security experts as the next Great Game of geopolitics, do not add up to a herd of elephants. To the disappointment of oil majors, there simply is not enough oil in the Caspian (compared to the enormous woolly-mammoth fields of the Middle East) to make a big difference in the global energy equation.

Just about the only virgin territory left is under the ocean. Underwater drilling is, in itself, nothing new. After all, the North Sea and near shores of the Gulf of Mexico have for decades been important oil-producing regions. Until recently, however, many geologists were convinced that offshore oil would be restricted to shallower waters. The sorts of rock conducive to oil accumulation, they argued, would be found only in ancient river deltas and other formations close to shore.

Veterans of the oil business recall that the notion of finding oil under thousands of feet of water was ridiculed until just a few years ago. Now oil majors are betting that enormous amounts of oil are trapped under the deep waters off Brazil, West Africa, and, of course, the Gulf of Mexico.

As you fly out by helicopter from New Orleans to Ursa, the entire history of America's offshore exploration unfolds below you in a scene as alive and crowded as any Brueghel painting. The shallow waters positively teem with activity as oil rigs, supply vessels, drill ships, and the like go about their business. The skies are abuzz with the sound of choppers ferrying crews and visitors to and from distant rigs, and hidden below the surface lies an intricate lattice of gas and oil pipelines that brings the precious hydrocarbons to shore safely and economically. James Dupree, head of deepwater production in the Gulf of Mexico for BP, points to an intensely busy map of the Gulf—with all the oil and gas pipelines carefully marked—hanging on his wall: "You see how all this activity tapers off by a depth of around 1500 feet? In a decade, a similar map will show that infrastructure going out to depths past 5000 feet."

Making that vision a reality is sure to be one of the biggest drivers of innovation in the oil business in coming years. Simply put, finding, drilling, and transporting hydrocarbons from ultra-deep water to market will be either uneconomic or downright impossible without further breakthroughs. Those few firms with deep enough pockets are taking the long view, and they see the billions of dollars required to develop such technologies as prudent investments. One Exxon technologist said that new techniques are needed "because we simply cannot afford to 'drill our way to knowledge' at $30 million to $40 million a pop per deepwater test well."

Though a century has passed since the Spindletop discovery, oil-production technology still remains more miss than hit. The average recovery rate across the globe is still a pitiful 30 to 35 percent: that is, of all the oil proved to exist in a given reservoir, companies typically can get barely a third to market. Technology that lifts recovery rates by even a few percentage points across a firm's portfolio is sure to translate into much bigger profits than the speculative, if sexy, business of hunting enormous "elephant" fields.

The key lies not only in cajoling more oil out of the main reservoir's oil-bearing rocks but also in tapping smaller (previously uneconomic) fields nearby using clever technologies like multidirectional wells. Euan Baird, who served for many years as the boss of Schlumberger, a French oil-services giant (rivaled only by America's Halliburton, which used to be run by Dick Cheney), has his sights set much higher: "We think we can develop technologies, particularly real-time monitoring of wells, that will help the industry lift recovery rates up to 50 or 60 percent within a decade."

Oil reservoirs are, in a sense, dwindling assets. Once companies start drilling, the reservoir is depleted constantly. Today's mature fields are declining by an average of 7 or 8 percent per year, and as much as 25 percent in places like Venezuela. The reasons for this vary from place to place, but they have to do with the dynamic interplay of water, natural gas, sand, and other forces "downhole," over which man has had little control—until now, that is.

Vast sums have been spent by the industry in recent years on such things as sand management, exotic cement coatings for wells, and other techniques designed to slow depletion. This problem will only get worse, since most of the world's oil fields are aging. In Britain's North Sea, for example, most large fields are now 70 to 90 percent depleted. Ironically, the same dramatic techniques that have allowed oil majors to improve oil-recovery rates (and so pump out more oil now) end up hastening the ultimate demise of those fields. As one boss puts it, "We are always running up a down escalator—and technology helps, but does not fundamentally change that."

The International Energy Agency forecasts that global production of oil must rise from around 80 million barrels a day today to some 115 million barrels a day in 2020 if anticipated demand is to be met. Industry experts at the Houston office of McKinsey, a management consultancy, argue that when that incremental demand is added to the already daunting task of compensating for depletion, the "technology challenge" amounts to adding a whopping 65 million barrels a day of oil production over the next two decades. By the IEA's reckoning, that means oil producers must invest $1 trillion upstream in non-OPEC countries over just the next decade—much of it, of course, in new technology.

The Technology Pipeline

"Innovation in the oil business has always happened from the bottom up, not the other way around. The roughnecks, not the guys in the research labs, usually deserve the credit!" Michel Halbouty was one of the last of the Texan wildcatters. He got his start in the business during the second Spindletop boom of the 1920s (the first boom kicked off by Higgins and Lucas at the turn of the century had, inevitably, ended in a spectacular bust that left so many investors holding worthless stock that the place was branded "Swindletop"). Halbouty was past ninety, but as sharp as the crease on his well-tailored suit. He remained a feisty maverick in a mega-merging world in which independent oil companies are getting to be an endangered species. Though he vowed never to sell out, even this stalwart son of Spindletop conceded that the future belongs to Big Oil: "The technology has gotten so sophisticated and so expensive now that only the majors can afford it."

So what sorts of revolutionary gizmos are these scientists at big firms coming up with? Actually, about the only thing they all agree upon is that there will be no single silver-bullet technology. Rather, there are likely to be a flurry of advances in three broad areas that together add up to improved recovery rates, lower costs, and so on: better visualization of reservoirs, better placement and drilling of wells, and—crucially—better management of those wells once they are in production.

The desire for better visualization of oil reservoirs is nothing new, of course. Back in the days of Spindletop, oilmen turned for help to "doodlebuggers"—soothsayers who claimed to be able to detect oil underfoot using a forked stick. As early as the 1920s, the industry started using very crude seismic analysis—essentially, setting off some dynamite in a hole to detect unusual patterns in the waves reflecting back to the surface.

The arrival of three-dimensional seismic imaging in the late 1980s and early 1990s transformed the industry. This technology has helped make sense of the earth's mysterious bowels, and the process of finding oil less hit and miss. By McKinsey's reckoning, the net benefit to the global oil industry from 3-D seismic imaging (through reduced drilling costs, additional reserves exploited, and so on) is about $11 billion a year. Today's visualization techniques still do not see through salt layers (such as are found under the deep waters of the Gulf of Mexico) very well. However, one Exxon researcher claims that his firm is developing "direct detection" techniques that will find hydrocarbons with 100 percent certainty. How would this magical technology work, exactly? Unsurprisingly, the firm is not willing to show its cards.

Topflight seismic data will surely lead to better placement and drilling of wells, but a number of other promising technologies could also emerge over the next decade. Already, today's drillers are working with technologies unimaginable to the last generation of roughnecks. On the Ursa platform, for example, the drillers who hit that amazing pocket of oil 28,000 feet away recently did so not while busting their guts or getting their fingers chopped off handling twirling pipe. Most of the dangerous work is now mechanized, and the supervision is done from a comfortable control room. Even so, said Tommy Morrison, the cheerful drilling supervisor on Ursa, it remains a risky business: "You just don't know you're gonna hit oil till you actually hit it!"

Even that might change in future. Much smarter drill bits, encasing sensors that measure the surrounding rock, would act as eyes and ears for the driller. By looking far enough ahead of the drill bit and communicating that information in real time back to the operator (or to a computer controlling the operation), the rig can be adjusted so that the bit safely hits oil every time.

Perhaps the most ambitious aim is better management of wells once in place. The use of chemicals with high pressure could enhance the fracturing of low-permeability formations, and so increase production. Installing compressors at the bottom of wells could fight declines in reservoir pressure over time, and so boost oil and gas recovery.

Another idea that some are now working on is "downhole processing." At the moment, firms often end up producing gas or, worse yet, water, when pure, sweet oil is what they are really after. Techniques for subsea separation of oil, gas, and water, using equipment embedded in the well itself or perhaps installed on the seafloor, could prove a dramatic improvement. In particular, if combined with new techniques for reinjecting unwanted water or gas back into the reservoir, this approach could prove to be a far cheaper and more productive way to extract oil.

The niftiest idea is that of next-generation wells that can essentially run themselves—or at a minimum call for human help when things go wrong. That is a far cry from today's dumb (and mute) wells, which typically lack instrumentation beyond crude pressure and temperature sensors. Researchers at Schlumberger's fancy research center near Cambridge, England, are working furiously on such developments. Locals call the complex the "circus tent" for its giant white canopy covering, which does strike visitors as a bit avant-garde for a nuts-and-bolts engineering laboratory. It turns out not to be an architectural affectation but rather a cleverly disguised safety feature: the superstrong canopy cover is designed to contain the damage in the unlikely event that any test well blows up. The clever young things cloistered away there and at Schlumberger's other global research centers are developing wells equipped with everything from elaborate electronic sensors to individual Internet addresses to communications links to local area networks. Their former boss, Euan Baird, draws inspiration from the medical world: "Today we manage oil wells as if the patient had to die before treatment, since wells cannot describe their symptoms to us. In future, real-time monitoring of wells will help alert us so that we can intervene and rescue the patient—or at least delay his demise."

The medical analogy is useful, but the exploration business will more likely resemble a different industry altogether: information technology. Roger Anderson of Columbia University is confident that "the wired, virtual oil patch will allow instant access to field monitoring of all the company's operations, and visualization over any laptop with proper access codes from anywhere in the world." Contrast that with the common practice today in many countries: some poor fellow drives around in a pickup truck from well to well once a day, jotting down the few basic measurements available by hand, and sends them off to the office sometime later to be tabulated by bean counters.

But as wells become more and more sophisticated, the oil industry risks being flooded with mountains of data. On Ursa alone, for example, Shell is already bombarded by 30,000 distinct data points every day. One of the chief challenges facing the oil industry in the next few years is developing the internal systems necessary to capitalize on this flood of information without being drowned by it.

Will the trouble be worth it? In some cases, such as well-understood onshore areas, undoubtedly not; a roughneck with a sharp pencil and a trusty pickup may be sufficient. But for the real plums—the riskier, greenfield projects offshore—the reward for designing the virtual oil well of the future from scratch could be handsome indeed.

Anderson advises the oil industry on how it can learn from the experiences of the automotive and aerospace industries in embracing new information technology tools and processes. "Those techniques have helped firms like Boeing [whose 777 airplane was designed 'virtually'] achieve amazing cost and cycle-time savings," he explained, "and they will help the oil industry slash its deepwater exploitation costs by forty percent over time." Add to that the vastly greater payoff possible in deep water (wells on platforms like Ursa are much bigger than typical onshore gushers) and you get a glimpse of the new economics of E&P technology.

That is sure to please shareholders of oil companies, but it's important to remember that Big Oil is still swimming against the tide of history. The sums that flow into oil exploration—and the future returns on that investment—will always hinge on the broader economic incentives facing the energy industry. The push toward liberalization of markets and the move to account for the environmental externalities of dirty energy are starting to level the energy playing field. As this trend gathers pace, powerful incumbents like oil and nuclear power (which, as the next chapter describes, has sucked up half of all the subsidies lavished on the energy industry in recent decades) will inevitably lose ground to newer, greener rivals.

Still, that transition to clean energy will not happen overnight. As the frenzy of innovation in E&P makes clear, the oil industry is unlikely to fade from the scene anytime soon. And while none of this fancy technology will change the locations on the planet where God put petroleum, it should certainly debunk the claims of those that insist that the world is about to start running out of oil.

10

A Renaissance for Nuclear Power?

MIDDLETOWN, PENNSYLVANIA, is the perfect antidote to the hustle and hassle of New York and Philadelphia. As you meander through the rolling farmlands of central Pennsylvania, friendly farmers tend to their crops and flocks, and local diners serve simple grub. Occasionally you come across an Amish horse and buggy. This pastoral setting seems a safe retreat from the complexity and perils of modern life. Until you spot an unmistakable specter on the horizon: the cooling towers of the Three Mile Island nuclear power plant.

Three Mile Island: the very words still send chills down the spine. It was in the early-morning silence of March 28, 1979, that one of the two reactors at TMI started to overheat. A combination of mechanical and human error sent the temperature in the reactor core soaring, threatening a blast that would have released unthinkable quantities of lethal radiation. Thousands of panicked locals began to flee. Politicians, TMI officials, and experts at the Nuclear Regulatory Commission bickered over what to do. Disaster was averted, but the world never forgot those five days.

Richard Thornburgh, who later went on to serve as attorney general under the elder George Bush, was then governor of Pennsylvania. He recalls the mayhem this way:

> I had little time to be personally frightened during the accident because of the constant press of the responsibility for the well being of nearly a quarter of a million central Pennsylvanians. The high points of my concerns were largely due to false or misleading information conveyed to the general public which required countermanding from my office. For example, the bogus evacuation recommendation from the NRC on Friday morning, March 30; the so-called "bubble" in the reactor reported on Saturday evening, March 31; and various news accounts exaggerating the potential for a nuclear meltdown throughout the incident. As a result of TMI, my level of skepticism about nuclear power was substantially raised and, like most Americans, I no longer took for granted the fact that this source of electric power was as risk-free as its promoters had indicated in its early years.

So many Americans shared Thornburgh's skepticism following the accident that the future of nuclear power seemed dark. In fact, not one new nuclear power plant has been built in the United States since then.

Its impact in Europe was great too. The TMI accident led directly to a national referendum in Sweden that demanded an end to nuclear power, though that phaseout has dragged on for much longer than envisioned. After the much more serious accident at Ukraine's Chernobyl nuclear plant in 1986, a number of other European countries turned against nuclear power as well. More recently, this nuclear autumn seems to have turned downright frosty. Asia had been the bright spot for the nuclear industry, but the Asian financial crisis of the late 1990s cooled that fervor. Taiwan, once a big fan of nuclear power, began to lose its enthusiasm after the long-ruling Kuomintang party was ousted from power. France, which remained defiantly pronuclear for years, also seemed to have lost its religious commitment to new nuclear energy: even François Roussely, the boss of Electricité de France, declared that he was "not wedded to nuclear power."

Ironically, the biggest blows to nuclear power were dealt by the industry itself, thanks to several relatively minor accidents that proved major public-relations disasters. In late 1999 it surfaced that British Nuclear Fuels (BNFL) had falsified records relating to a shipment of nuclear fuel to Japan, sparking outrage in both countries. The Japanese insisted on sending the shipment back to Britain—which created a huge financial mess for BNFL as well as a great public-relations coup for Greenpeace protestors. Making matters worse were revelations that the firm had understated the cost of nuclear cleanups in Britain by more than $12 billion. In parallel, a similar farce was playing in Japan: shoddy management practices at an experimental fuel-reprocessing plant in Tokaimura led to the deaths of two workers after they were exposed to radiation more than 10,000 times the level considered safe.

These reckless blunders fueled a backlash in Germany, where environmental groups succeeded in their long campaign to halt nuclear power. The government of Gerhard Schroeder decided to end reprocessing of nuclear fuel by mid-2005—another big financial blow for BNFL. Germany and Belgium both passed legislation banning new nuclear plants, though the political compromises struck with the industry will allow existing plants to run out their useful life. In Japan, the government quietly scaled back its plans for twenty new plants and, in 2002, started forcing the country's largest utility to shut down its nuclear plants temporarily over yet another scandal involving falsified safety data.

A New Dawn for Nuclear Power?

Yet there are still some believers in the technology that early proponents claimed would produce electricity that would be "too cheap to meter." In 2002, South Africa's Eskom, the local power utility, was working with the beleaguered BNFL to develop the next generation of nimble, supersafe reactors using something called "pebble bed" technology. Finland's TVO, a power company controlled by a consortium of heavy energy users, persuaded local officials that it made sense to build a $2 billion nuclear plant near Helsinki. In America, the Tennessee Valley Authority, a federally owned utility, also tried to revive nuclear power.

Some politicians began to soften their line. Loyola de Palacio, the top energy official at the European Commission, stuck her neck out for the industry by arguing that European leaders faced a hard choice—they could shut down nuclear plants immediately to please radical green groups or they could meet commitments to reduce emissions of greenhouse gases under the Kyoto Protocol. But, she insisted, they couldn't do both: "Will someone explain to me how we [can] substitute nuclear energy when it represents more than 15 percent of the EU's energy supply? I myself have proposed a push toward renewable energy, but it isn't enough to cover the share from nuclear. I have also proposed projects to improve energy efficiency, but it continues to be insufficient. I don't have a particular fondness for nuclear energy . . . [but] Europe cannot do without nuclear energy unless it abandons the objectives of Kyoto. Each year nuclear saves 300 million tons of CO_2 in the EU—three times more than everything we have to save over the next decade to comply with Kyoto." She knew that nuclear power, which emits no greenhouse gases, would probably be replaced by dirty fossil fuels if plants were shut down overnight.

On the other side of the Atlantic, a cabinet-level task force unveiled President Bush's much-awaited energy policy. The headlines at the time focused on the plan's fears of an energy "crisis" and its controversial call for throwing open the wilds of Alaska to oil drilling. In fact, the most enduring legacy of the plan may prove to be its endorsement of nuclear energy. Vice President Dick Cheney, the head of the task force, argued forcefully not only for giving existing nuclear plants a renewed lease on life, but also for building a number of new plants.

As these developments suggest, nuclear energy just might have a chance to return from the dead. The European Nuclear Society and the European Atomic Forum, big industry lobbies, organized a conference that captured the new mood: "Nuclear Energy Sector Ready to Show its Strength: ENC 2002, the European Nuclear Conference taking place in Lille, France, on 6–11 October, will provide visible proof of the vitality and 'staying power' of the nuclear energy sector in Europe and other parts of the world."

Although the industry's prospects are not as dark as they were in the aftermath of the Three Mile Island or Chernobyl accidents, industry talk of a renaissance is much too rosy. What is fair to say is that after many years of ignoring nuclear power altogether, policymakers are now starting to engage the issues once again. That does not mean, however, that the debate over nuclear power is healthy or productive. Peter Beck and Malcolm Grimston, experts at London's Royal Institute of International Affairs, describe the tussle as the most cantankerous in the energy field:

> [Nuclear power's] supporters are confident that [it] will have an important long-term future on the global energy scene, while its critics are equally confident that its days are numbered and that it was only developed to provide a political fig-leaf for a nuclear weapons program. Both sides believe the other to be thoroughly biased or stupid and there is little constructive debate between them. As the disputes rage, especially over such issues as the management of nuclear waste, the economics and safety of nuclear power compared with other sources of electricity, the possible links with nuclear weapons and the attitude of the public towards the industry, decision-making is either paralyzed or dominated by those who shout loudest.

In short, policymakers are caught in a nuclear bind.

On one hand, the industry is still beset by worries about the safety of its plants, the lack of storage facilities for nuclear waste, and its sorry history of cost overruns. On the other, its safety record has improved dramatically in recent years, plans for dealing with waste are emerging, and operators claim to be making good money from nukes these days. Many of the world's reactors are aging and will need to be replaced in the next couple of decades—but with what? Renewable energy, demand the greens. New nukes, insists the industry. Probably fossil fuels, say the cynics. To judge whether a nuclear revival is really in the cards, it's worth looking closely at why exactly the industry is so cheerful today.

Cranking Out the Juice

A good place to start is back at Three Mile Island itself. The near-tragic accident at TMI two decades ago wiped out one of the plant's two reactors, but the remaining reactor has been back in service for some years now and is managed by a division of Exelon (a giant American utility that has been busy gobbling up nuclear power plants). In that time, Three Mile Island has become one of the most efficient and safest nuclear plants in the country; it is also one of the most profitable. Corbin McNeill, who was Exelon's boss in 2001, believes that financial success and safety are, in fact, intimately linked. He saw in TMI a shining symbol of nuclear power's bright future.

Global liberalization of energy markets can be thanked for this turnaround. Nuclear plants are finally beginning to be run as proper businesses by competent, serious managers rather than by incompetent local monopolies. Peter Beck and Malcolm Grimston explain that the impact of liberalization on nuclear power has been particularly significant because this coddled corner of the utility business had been extremely inefficient.

The effects are best seen in America, where wholesale markets for power were deregulated in 1996, though the nuclear industry elsewhere is following similar trends. The result of deregulation was a painful squeeze on America's dozens of nuclear plants, many of which were run as one-shot investments by local utilities. That is rapidly changing, however, thanks to a flurry of mega-mergers (like the one that created Exelon), joint ventures, and other sorts of management consortia. Experts think the trend will continue.

Consolidation makes sense. Plant managers benefit from economies of scale in fuel purchases, maintenance crews, and so on; they can also more easily apply best practices among various plants. The results have been striking, in terms of both rising capacity utilization and diminished waste of fuel. The operators of existing nuclear plants have also managed to boost the output of their existing plants by upgrading their steam generators and turbines. As a result, America's nuclear plants cranked out power during the winter of 2000–2001 at an operating cost of just 1.8 cents per kilowatt-hour. Coal plants produced it for 2.1 cents per kW-hour, while those using natural gas (the price of which soared that winter after California's power crisis) managed only 3.5 cents per kW-hour.

Such improvements, argue nuclear advocates, make a clear case for extending the licenses of existing nuclear power plants beyond the original limit of forty years. A number of plant permits will start expiring in America come 2006, and nearly all will do so by 2030. In many parts of the developed world, the picture is similar. The improved safety has in fact convinced the neighbors of nuclear plants that extending licenses is desirable. Several plants have already received approvals for another twenty years of operation from nuclear regulators; more will soon.

Cheap Today, Cheap Forever?

Here's how devotees of nuclear power see things. The cost-competitiveness of many existing nuclear plants is but a harbinger of the altogether "new economics" of nuclear power that will pave the way for good times to come. These devotees point to promising new designs (such as that "pebble bed" design that intrigued South Africa's Eskom) and suggest that new power plants will be safer and cheaper than today's. They also argue that costs will be much lower in the future because of the maturation of the industry: both companies and regulators, they think, now know how to avoid the needlessly expensive bureaucratic quagmire that followed the TMI accident. All this adds up to new plants that, in the words of vendors, will be cheaper than coal plants.

Don't believe them. Even if the new designs do prove safer in practice, they may not necessarily be cheaper. By the calculations of the International Energy Agency, the capital cost for these new nuclear designs runs about $2,000 per kilowatt electricity, versus about $1,000 to $1,200 for coal and just $500 for an efficient natural-gas plant. The cost of power from future nuclear plants will therefore likely be much higher than the two cents per kW-hour they appear to be from today's plants (whose debts are largely written off). In fact, the true cost of power from today's plants is at least double the apparent figure, once various government subsidies and "externalities" are accounted for.

Capital cost is a critical hurdle for nuclear power. When measured on a present-value basis, some 60 to 75 percent of a nuclear plant's costs may be incurred at the beginning of its life; for a gas plant that figure may be just 25 percent. What makes these costs an even more awkward starting point for nuclear power is that they do not include interest costs accrued during construction. Given the many years it typically takes to build nuclear plants, that can make or break a project. Their capital-intensive nature also makes nuclear projects exceptionally sensitive to the cost of capital.

The nuclear industry would tell you to ignore today's costs, since tomorrow's plants, benefiting from lessons learned from decades of hard knocks, will prove cheaper to operate. They say that they will build bigger plants and that they will take advantage of economies of scale by building a series of identical plants at once, rather than uneconomic, one-shot structures as they have in the past.

Unfortunately, building bigger plants introduces greater complexity, which inevitably introduces greater uncertainty and cost. The idea of achieving scale economies by building many plants is thwarted by an inconvenient fact: the electricity market in the developed world is simply not growing fast enough to need lots of giant power plants, no matter the type. After all, the clear trend in deregulated electricity markets is toward smaller plants; it is micropower, not megapower, that the market favors these days, thanks to the far lower financial risk involved.

There is something to the argument (often voiced by company bosses) that a more mature industry will mean smarter businessmen and regulators alike. Undoubtedly, the reaction worldwide to the TMI accident resulted in many needless, overcautious regulations. These in turn did delay nuclear projects, especially in America: while Japan's government has managed to ram through nuclear plants in five years or so, some American projects have taken ten or fifteen years to complete.

In one infamous case in New York, the Shoreham nuclear power plant was completed, but grassroots activism kept it from ever running. Local officials thumbed their noses at the feds and refused to implement an evacuation strategy for the plant's neighbors. LIHistory.com, a website that chronicles developments in Long Island, New York, describes the struggle in heroic terms: "Shoreham launched an anti-authoritarian brand of citizen activism that transformed local politics, especially in Suffolk County, and against all odds vanquished the massed power of the federal government, Wall Street and the electric utility industry, preventing a completed and fully licensed nuclear power plant from operating for the only time in American history." By 1994 the Shoreham plant was dead but hardly forgotten: its $6 billion cost left locals paying some of America's highest electricity rates. Regulators and local officials are no longer that obstructionist.

Even so, the real cause of delays in nuclear development was probably the inadequacy of the technology—not needless red tape. Many new plants happened to come on-line at around the time of the TMI accident, and a number had generic technical problems that deserved attention anyway. Even the French program, touted as a model of technical and scale efficiencies, suffered from several such problems as late as the 1990s: its latest N4 design, for example, was found to have cracking problems in its safety-related heat removal systems. The IEA analysis insists that the "cost studies are, to a greater degree than at any time since the 1970s, dependent on paper designs that have been demonstrated only in a few countries."

Clean and Green—or Obscene?

The splashy full-color advertisements seemed to be everywhere in early 2001: perky kids bobbing to their Walkman stereos, smiling and playing and saying things like "Clean air is so twenty-first century!" You could have been forgiven for thinking the campaign was designed by Greenpeace or Friends of the Earth to attract high schoolers to their cause. In fact, it was bankrolled by the nuclear power industry as part of its stealthy campaign to steal the green high ground. After decades of being pilloried by environmentalists, the industry thought it had finally found a way to attack its opponents' Achilles' heel: global warming. Because nuclear power emits virtually no greenhouse gases, nuclear lobbyists reasoned, it deserves to be applauded as an environmentally friendly energy source.

That clever bit of logic ignores one inconvenient fact: the industry still produces waste that is the most toxic, environmentally unfriendly stuff ever invented by man. Even so, that ad campaign makes clear the industry's strategy for survival. Once free-market arguments are exhausted, its advocates will point to other benefits of nuclear power that they reckon are worth paying for: "security of supply," environmental benefits, and the like. In some countries and in some circumstances, such arguments might have some merit. But is nuclear energy really a special case that deserves subsidies or other sorts of government intervention?

The arguments on energy security vary, but they boil down to diminished reliance on fossil fuels, less vulnerability to embargo from OPEC, and reduced import bills. They are powerful but misleading. OPEC does control the oil market—but oil is not used much in power generation these days, just for transportation. Nuclear power, in contrast, is not used at all for transportation. Whatever the political merits of diversifying energy supplies, an analysis done in 1998 by several agencies affiliated with the OECD is quite clear in its assessment of their economic merit (even after accounting for amortization of expenses over the life of the plants): "For many countries, the additional energy security obtained from investing in non–fossil fuelled generation options is likely to be worth less than the cost of obtaining that security."

What's more, it is surely perverse to talk about nuclear power as enhancing energy security post–September 11. For a start, creating lots of fissile material that might be stolen by terrorists or rogue states is an odd way to look for security. Also, one important lesson from those terrorist attacks was that high-profile infrastructure is extremely vulnerable. And although the terrorists behind the attacks on the World Trade Center chose not to target nuclear plants, the next bunch might very well want the extra bang for the buck that just possibly might come from hitting a nuclear power plant. Paul Leventhal of the Nuclear Control Institute, a watchdog group based in Washington, D.C., insists that an attack could release "a plume of radioactive materials, particularly over a nearby city, [and] would dwarf the consequences of the World Trade Center and the Pentagon attacks." His group's study of the matter suggests that perhaps "tens of thousands" of cancer deaths would result. The industry, predictably, rejected such claims as nonsense and insisted that security has been beefed up post–September 11.

However, the experts at the International Atomic Energy Agency suggest that there may be cause for concern after all: while it thinks a jumbo jet crashing into a nuke would not trigger a runaway nuclear reaction, thanks to built-in safety systems, it did think that a breach of the containment vessel and a leak of radioactive steam and fallout were possible. A government report looking at the vulnerability of Indian Point, a nuclear plant near New York

City, added to such concerns. *The New York Times* ran a front-page story describing the report as painting "a picture of potential chaos brought on by panicked parents rushing to pick up children from schools, firefighters unsure what to do and antiquated computer technology hampering predictions of where the radiation might be headed and how many people would be at risk." Not much seems to have changed since Three Mile Island.

On environmental grounds, too, nuclear does not come out a clear winner. It is true that nuclear energy does not produce carbon dioxide; that, say fans like America's Vice President Cheney, means that the world "ought to" build more such plants. Really? How? Handing out government money (through production or investment tax credits, for example) is a demonstrably murky, inefficient way for government to encourage the climate-change benefits of nukes.

The straightforward way would be through some sort of carbon tax, which would penalize fossil fuels but not energy sources free of carbon emissions. The IEA's experts have analyzed how much of a boost nuclear power could get from a carbon tax, and the answer is not the clear-cut win for nuclear some expect. This sort of crude calculation is wide open to second-guessing, but assume any initial carbon tax falls between $25 and $85 per metric ton of carbon, the range many experts think may be needed to get developed countries to their Kyoto targets. The IEA thinks that even the highest of those taxes would boost the competitiveness of nuclear electricity versus coal (which is carbon-intensive) by just two cents per kW-hour, and versus natural gas by just one cent per kW-hour. At the low end, nuclear power benefits by just half a cent and a quarter cent, respectively.

At the end of their deliberations on the future of nuclear power, experts at the Royal Institute of International Affairs reached this sobering conclusion: "A clear prerequisite for a nuclear renaissance seems to be that the economics of nuclear power would have to improve beyond those of the currently deployed technology, which is based largely on designs from the 1960s and 1970s. Factors such as climate-change policy and fears about the security of energy supplies may be of marginal benefit to nuclear power, but by themselves are unlikely to make the crucial difference."

This is especially true when one considers the grave environmental liabilities that nuclear energy itself carries. Nuclear radiation is a threat throughout the production process, from uranium mining to plant operation to waste disposal. During that entire cycle, there is that unthinkable if tiny risk of an accident that would seriously harm the environment and human health. And waste disposal, despite decades of research and consultations and politicking, remains a farce: not one country in the world has built a permanent waste disposal site. The United States hopes to have one finished at Yucca Mountain, in the Nevada desert, in a decade's time; the European countries are a decade behind that.

Even if these geological storage sites are completed, they are no real solution to waste that will remain deadly for perhaps 100,000 years. The best that the world's sharpest nuclear minds could come up with after fifty years of research and endless pots of money is to take this horrid stuff, bury it in a big hole in the ground, and pray that our grandchildren will be clever enough to figure out how to make it safe.

The Invisible Hand Torch

The above approach to the waste problem points to the one aspect in which nuclear energy does deserve to think of itself as special: subsidies. While many sorts of energy receive some government boost, especially during their fledgling days, nuclear is unique in the scope, scale, and stealth of subsidies received. Though the particulars vary from country to country, the range of financial assistance offered to the industry from taxpayers' wallets is astonishing: subsidized fuel and fuel processing; research and development funds; capital at artificially low rates of return; subsidized waste cleanup and disposal; and so on.

How much does all this add up to? Reliable, comprehensive, and up-to-date global figures from neutral analysts are nowhere to be found. The industry itself usually maintains the outrageous fiction that it no longer receives any subsidies at all. The American industry's official position is that there is absolutely no subsidy involved in the Price-Anderson Act, through which Congress limits the civilian nuclear industry's liability for nuclear catastrophes to only $9 billion (a small fraction of what a Chernobyl-scale disaster would cost in America). Estimates by nuclear critics like Greenpeace and Friends of the Earth carry too many zeroes to fit on this page.

According to official figures, OECD governments poured more than $150 billion in today's money into nuclear research during the last quarter of the twentieth century. That breathtaking subsidy accounted for more than half of all money spent by these governments on energy research in that time—yet the industry now demands even more. One has to wonder how much could have been achieved if that money had instead been invested in, say, renewable energy or fuel cells.

In the end, nuclear energy's future may be skewered by the same sword that is making it fashionable today: the deregulation of electricity markets. This tidal wave of reform spreading through the world's moribund energy markets, combined with the current volatile price of fossil fuels, makes existing nuclear plants appear exceptionally attractive. But that is true only for old plants that are mostly paid for or have gone bankrupt; some, like TMI, have been bought for a pittance and so crank out power for virtually nothing.

The same process of liberalization that made such fire sales inevitable is also exposing the true economics of new plants. As subsidies fade, the industry will once again be brought down to earth. And TMI may yet come to be seen not as the bright new hope for the nuclear future, but rather as a symbol of the human tragedy that nearly was—and, if politicians do not pay attention, a testament to the economic folly that may yet be.

If the industry were to put its own money into research and development (and governments should ignore activists and encourage such research), its rocket scientists just might be able to come up with some amazing breakthroughs that could overcome the objections laid out above. Exelon's Corbin McNeill, for one, was convinced that the energy future does indeed belong to hydrogen-powered fuel cells. However, he argued that tomorrow's breakthrough nuclear power plants could be the source of the electricity needed to break down water to produce that hydrogen fuel.

Sadly, there is little sign that the nuclear industry has either the foresight or the deep pockets required to invest in long-term research—even with government assistance. Throughout the world, in fact, funding for nuclear research at universities and at corporate laboratories has dried up in recent years. The number of new doctoral students has slowed to a trickle, and the industry's average worker is getting geriatric: in the United States, three out of ten nuclear engineers are expected to retire within the decade. These are hardly the signs of an industry that deserves—or even believes it deserves—to be revived. On the contrary, it is an industry that seems destined to slip, in the words of early opponents (including Albert Einstein), "into ever less capable hands."

One overcast afternoon I met Walt Patterson at the French House, a legendary bar and restaurant in London's Soho that was a watering hole for the French Resistance in World War II. As he savored a delectably rare cut of beef, he told me the tale of how a small cabal, which included the Sage of Snowmass and the Archdruid, launched the campaign that eventually brought down Britain's nuclear establishment. Back in the 1970s, he explained, the country's environmental activists were nothing like the professional, well-funded machine that operates today. In those days, a young Patterson joined hands with such iconic figures as Amory Lovins (the energy guru who makes his home on a mountaintop in Old Snowmass, Colorado) and David Brower (the founder of Friends of the Earth, who was immortalized by John McPhee in his book *Encounters with the Archdruid*) to fight the nuclear industry. Against all odds, the campaign succeeded: "When we started, the industry was planning at least thirty new pressurized-water reactors in Britain by 1982; in reality, after all these years, they have built just one of those—the one at Sizewell."

Patterson is astonished at renewed talk of nuclear power, this time as a carbon-free energy source: "I guess we never quite drove a stake through its heart." He is encouraged by the political and media savvy of the new generation of nuclear opponents, but he observes with a touch of sadness that they sometimes reinvent the wheel: "They tend to forget history and despair too easily. We've fought these battles before, and we won. The next time the industry makes grand promises about the future, just ask them one simple question: When have you ever delivered a plant on time and on budget, that worked to its original specifications?" What really upsets him about all that wasted money is not the nuclear plants themselves, risky as they are, but all of the much better things—especially energy efficiency—that those many billions could easily have bought instead.

That reverie from one of the grizzled veterans of La Resistance Nucléaire points to the most powerful reason to think of nuclear power as a failed technology: economics. Not long after Patterson uttered those angry words, the nonsensical economics of the industry were exposed once more—as was the willingness of government to throw taxpayer money at it. British Energy—a privatized company that runs most of Britain's nuclear power plants—declared itself out of money and demanded government handouts. Britain's liberalized electricity market, it said, was too competitive; the firm's nukes simply couldn't produce electricity cheap enough. Market-minded experts argued that the firm should be allowed to go bust. Citing the usual arguments about security of supply and climate change, the industry and its friends clamored for a bailout. The government caved in. And the ordinary taxpayer paid the price for this folly, yet again.

Back in 1956, during the dizzyingly hopeful early days of the industry, John von Neumann of America's Atomic Energy Commission had predicted that a "few decades hence, energy might be free, just like the unmetered air." History has clearly proved him wrong.

Concerns about climate change might provide this industry a reprieve for some time in some parts of the world, but the broader trend is clear: big nukes, like giant hydroelectric dams (described in the next chapter on energy in the developing world), are being swept aside by the global move toward micropower.

Barring some extraordinary technological breakthrough, the industry that once boasted it would be too cheap to meter is likely to be remembered as too costly to matter.

11

Micropower Meets Village Power

BILL GATES and Gerardo Zepeda Bermudez do not see eye to eye. One of them thinks that technology, especially fancy new stuff like high-speed Internet access, is the key to alleviating poverty. The other finds such views hopelessly naïve. They epitomize the polarized debate over how to bring the benefits of modern energy to those around the world who languish in poverty without reliable access to electricity or safe fuels.

The technology enthusiast points to San Ramón, a hardscrabble village in a remote corner of Honduras that was devastated by Hurricane Mitch a few years ago. In rebuilding the area, which had no access to electricity before the storm, officials decided "to do it smartly," according to the technophile: they installed solar panels and used them to light up common areas such as the rebuilt school and community hall. They also used the power to fire up a dozen personal computers with fast Internet access, as well as for videoconferencing facilities. The project's proponent believes that such "solar net villages" are the wave of the future: "We don't see this as just an energy project, but rather as a poverty alleviation project."

Humbug, says the other man. He insists that the Internet will, in itself, do nothing to address the real concerns of the world's poorest. At a gathering devoted to using technology to bridge the "digital divide" between rich and poor, this fellow proved the skunk at the garden party. "Do people have a clear view of what it means to make a dollar a day?" he asked, referring to the threshold used by economists to measure "absolute poverty," which means the lack of even basic nutrition needed to stay healthy. Perhaps a billion people live in such squalor today. He added, "There's no electricity in that house. None!" Even in the face of San Ramón's inspiring technological leap, he was steadfast: "You can't afford a solar power system for less than a dollar a day; you're just buying food, you're just trying to stay alive."

It is no surprise that views differ sharply on such a difficult topic. What is a surprise is that the techno-optimist is not Gates, the founder of Microsoft and the world's richest man. It is Zepeda, who was a minister in the Honduran government that rebuilt San Ramón. Bridging their opposing visions is a powerful argument for rural energy that is now taking hold among energy thinkers at development banks, nongovernmental organizations, and in the private sector: off-grid, community-based distributed generation. Or more snappily put, Village Power.

The question of how best to help the "energy-poor" is not merely an arcane debate among technology geeks and development economists: it is one of the most pressing global dilemmas of the new century. At the big Earth Summit II held in Johannesburg, South Africa, in 2002, world leaders declared that tackling energy poverty would be at the top of the list of things to do to ensure sustainable development in the future.

There is a moral imperative at work that justifies their sense of urgency. In an age of unprecedented global prosperity, it is outrageous that so many unfortunate people should continue to live in such grinding poverty. Another, more selfish, reason for the rich world to care is that the energy needs of the world's poor will—unless governments intervene—increasingly threaten oil resources, international trade, and global economic growth. The well-being of the world's energy-rich in Manhattan and Munich looks to be increasingly and inexorably linked to the fate of the energy-poor in places like Mali and Manchuria.

Most people in the rich world have never seen the true conditions in which the world's poorest live. Even tourists in places like Jamaica or Zimbabwe generally see only the urban poor or those who live near tourist enclaves, and they are usually better off than their rural kin. While those lucky enough to live in rich countries benefit from all of the mobility, warmth, and economic productivity made possible by ready access to modern energy, much of the world still uses energy more or less as mankind did thousands of years ago.

Living in the Stone Age

The World Energy Assessment (WEA), a comprehensive effort by the United Nations and the World Energy Council to help shape the future of energy, opened its final report with these words:

> One way of looking at human development is in terms of the choices and opportunities available to individuals. Energy can dramatically widen these choices. Simply harnessing oxen, for example, multiplied the power available to a human being by a factor of 10. The invention of the vertical waterwheel increased productivity by another factor of 6; the steam engine increased it by yet another order of magnitude . . . In industrialized countries people use more than 100 times as much energy, on a per capita basis, as humans did before they learned to exploit the energy potential of fire.

In a world of plenty, though, there is still great deprivation. The report's authors go on to warn that "the current energy system is not sufficiently reliable or affordable to support widespread economic growth. The productivity of one-third of the world's people is compromised by lack of access to commercial energy, and perhaps another third suffer economic hardship and insecurity due to unreliable energy supplies." In other words, it is not just lack of money that keeps the poor mired in poverty—it is also the lack of access to energy and all of the essential things made possible by it.

The International Energy Agency estimates that perhaps 1.6 billion people are without access to modern energy. Even the ambitious investments now planned—which assume that more than $2 trillion will be spent to upgrade the poor world's electricity sector by 2030—will still leave 1.4 billion people without access to modern energy in three decades' time. In other words, unless the rich world redoubles efforts to help those whom Bill Gates got so upset about, too many will still live entirely outside the modern economy. That would surely be a tragedy.

Just because the indigent are energy-poor, however, does not mean they use no energy: life would be impossible without it. What it does mean is that they use mostly noncommercial fuels, such as charcoal, crop residues, and cow dung, usually in ways that are harmful to both human health and the environment. Anyone who has traveled on India's roads, for example, knows that wandering cows are not the only hazard facing drivers. So too is the sudden appearance of hundreds of cow patties, neatly arranged on one half of the road by a local villager drying them in the sun so they can be sold as fuel. Out of necessity, many poor people also chop down forests for firewood and to make charcoal. Consider the plight of impoverished Zambia: since wood meets the cooking and heating needs of more than 80 percent of its population, precious hardwood trees are being cleared at an alarming rate. As a result, Zambia has one of the worst deforestation problems in the world.

Such inferior fuels make up a whopping 25 percent of the world's overall energy consumption and 75 percent of all energy used by households in developing countries. Even in the urban areas of poor countries, which have greater access to liquid and gaseous energy, only a small circle of elites enjoys high-quality energy; even when the masses in the slums use the same fuels, they use them in cheaper, dirtier stoves and boilers.

The environmental consequences of dirty energy use are bad enough, but more appalling is the toll on human life and economic productivity. The WEA estimates that 2.5 million premature deaths a year worldwide are due to exposure to indoor air pollution caused by burning solid fuels in poorly ventilated spaces. As the earlier chapter on air pollution explained, women and girls are hit particularly hard, since they typically spend many hours a day fetching fuels and enduring noxious fumes while cooking on makeshift stoves.

The Tata Energy Research Institute (TERI) in New Delhi has done an enormous amount of work in the area of indoor pollution. Its director, Raj Pachauri (who is also the head of the UN's Intergovernmental Panel on Climate Change), is convinced that modern ways of using fuel can help the poorest women improve their lives. To that end, his group has been developing and distributing simple but modern cookstoves that burn even primitive fuels cleanly. Electricity can have a dramatic impact on education and literacy, for children can study at night after they have completed their chores. Pachauri sums it up this way: "If the world is to have any hope of lifting the yoke of absolute poverty, it must provide rural villages and urban slums with far greater access to electricity and modern fuels."

More for You, Less for Me?

Energy consumption in the developing world has long been dwarfed by consumption in the rich world, both in absolute terms and in terms of use per person. However, that could change dramatically over the coming decades as Asian economies, especially India and China, grow wealthier, more urban, and more likely to use commercial fuels. Over the past thirty years, the commercial energy use of developing countries has grown three and a half times faster than that of rich countries; and more than two-thirds of the increase in energy demand from 1997 to 2020 is expected to come from the developing world. This trend could have a serious effect on the availability, cleanliness, and maybe even stability of the world's energy supply.

If Asia's rising economies use energy in ways that are as inefficient and polluting as, say, the United States, then the rich world is set for a rude awakening. "The problem is compounding," explains Exxon's Lee Raymond. "The 8 percent annual growth in oil demand from developing countries was easy to meet when those economies were smaller." However, he goes on to argue, as those economies grow bigger, that sort of growth rate will translate into a mind-boggling number of extra barrels of oil that must be produced by someone, somewhere.

As Asian economies demand ever more oil, they will add to the market power and political clout of a handful of OPEC producers in the Middle East, where the lion's share of proven oil reserves lie. Indeed, the relationship between China and Saudi Arabia could, in the long run, prove as important in geopolitical terms as the one between China and the United States. Thanks to roaring domestic demand, China went from being a net energy exporter to a net importer in just ten years. That has led to growing unease among China's leaders about becoming overly reliant on imported energy, and has unleashed a frenzied quest for energy self-sufficiency. Some officials and academics are convinced that the country's long-term energy future lies with a shift to a fuel-cell economy based on domestically produced hydrogen.

If a shift to superclean hydrogen energy really was to happen, it would be good news for everyone everywhere (except perhaps OPEC countries). Unfortunately, that day is probably some way off. Most "business as usual" scenarios predict that China will continue to rely heavily on dirty technology to produce power from its plentiful supplies of coal. If China and India do take that polluting path, they will surpass even the United States as the leading emitters of carbon dioxide. That, warns David Hawkins of the Natural Resources Defense Council, would cast a grim "carbon shadow over the rest of the century." It would also make any efforts by rich countries to curb global warming, such as the Kyoto Protocol, irrelevant.

Given the sums involved and the very long life of assets in the power business, this is no trivial matter. The least expensive (in the short term, anyway) path for the leaders of the developing world is to follow the fossil-fuel trail blazed by the rich world. After all, why should poor countries pay extra for cleaner energy when the richest country in the world still gets half its electricity from dirty and inefficient coal plants that are decades old?

We are at a historic turning point. Within a few decades, the energy aspirations of the developing world will move from the margins to front and center on the world stage. Unless rich countries help poor countries to leapfrog over the dirtiest energy phases that the rich world went through, the future for all may be needlessly grim.

What are the chances of such a leap? China is indeed trying to learn from the experience of rich countries, adopting clean-emissions technologies and regulatory innovations like emissions trading. However, it is the exception rather than the rule. If the experience of the poor world is your only guide, you would have to conclude that the outlook is bleak. Rather than help the poor obtain clean energy technologies and best practices, the developed world has consistently—and some would add cynically—exported dirty technologies and worst practices to the poor world. Some groups have long lobbied for OECD countries to "green" the way they do export finance. At the moment, agencies such as Japan's Bank for International Cooperation do just the opposite: the lion's share of loan guarantees, risk insurance, and other de facto subsidies lavished on the energy industry go not to renewables, but to fossil fuels. Cresencia Maurer of the World Resources Institute (WRI), echoing the sentiments of many angry campaigners, asks how supporting fossil fuels is consistent with the obligation of rich countries to transfer clean technologies to poor countries under the Kyoto treaty on climate change. Never mind the more basic question of why such mature, well-established industries should receive even a penny of taxpayer help to peddle their wares in the developing world.

Shameful as such rich-world practices are, though, the biggest culprits are the failed policies of central planners and cosseted bureaucrats in the energy and development ministries of poor countries. The world's energy poverty has more to do with the failings of governments and institutions than with genuine scarcity of electricity or even money, as a close look at the politics of power in India will show you.

Fixing the Leaky Bucket

You might think that electricity is one thing you could rely on in Bangalore. The Indian city is a much-celebrated center for software development that generates billions of dollars of wealth for its state, Karnataka. It has produced such global success stories as Infosys and Wipro, firms that are well known and respected on Wall Street and Silicon Valley and that have turned local paupers into princes. Bangalore's golden goose depends entirely on a steady, sure supply of electricity reaching the software firms and technology parks, so it reasonable to assume there is reliable power in the city, right?

Wrong. In fact, power outages are all too common. Many local businesses have given up altogether on the Karnataka State Electricity Board (KSEB), the local monopoly provider, and are instead setting up their own power plants, many fed by dirty diesel fuel. Most of India's state electricity boards are no better than the KSEB: they too deliver unreliable power to consumers and massive deficits to their government owners.

When I was in Bangalore, much of the city was recovering from yet another power outage. I was visiting N. Gokulram, then the boss of the KSEB, and when I asked how much such system failures cost the local economy, he threw his arms in the air in despair. He sheepishly admitted that he had fielded angry calls the previous night from politicians and journalists while sitting in the dark himself. Part of the problem is shoddy infrastructure, but the board itself is worse. Reluctant to cut people from the grid when a problem arises, the agency often spreads too little power among too many users. As a result, everybody receives poor-quality electricity. Consumers pay for this too, in ruined motors and machinery, lost output, and crops denied irrigation. India, not surprisingly, is the world's biggest market for such oddities as power stabilizers and voltage "correctors."

Even with such shoddy service, the KSEB, like all the state electricity boards, was still bleeding red ink (and had to be subsidized by the state government budget). Across India, the subsidies required by SEBs topped $13 billion in 1999, a whopping 3.5 percent of India's GDP. Electricity supply was a main reason for the state governments' budget deficits.

Although all consumers pay indirectly for power, few pay in cash. Barely more than a third of Karnataka's total power consumption was metered. Farmers, a politically significant group, received unmetered power at absurdly low flat rates that simply encouraged waste. Many industrial users, fed up with cross-subsidizing the deadbeats, switched over to captive power plants; their departure from the grid left fewer to pay the bill. Worse still, the KSEB's transmission and distribution losses amounted to about 30 percent, three times the international norm. Much of that was outright theft. All this was a recipe for disaster.

The curious case of Karnataka explained why India was the most perverse market for electricity in the world. Although the country produced more than 115,000 megawatts of electricity, its consumption per person remained tiny. And despite the fact that more than 15 percent of all India's capital investment in the 1990s went into electricity, the country still suffered a power deficit at peak hours of as much as 20 percent. If this sorry state were a function of only poverty and overpopulation, it would be sad but understandable. Yet even China (with a bigger population and greater geographic spread) delivers twice as much power, on average, to its population. In fact, one of the unsung achievements of the last few decades is the doubling of the number of households with access to electricity in China. Indians are denied greater access to electricity, and so to the necessities and comforts of life made possible by it, because of the incompetence of their political masters.

To be fair, India's government has tried to come to grips with this problem for much of the past decade. But rather than tackling the task of "fixing the holes in the leaky bucket," as one observer said in describing the SEB morass, the federal government first pursued the easier path of opening the taps to pour lots more power in. Early reformers tried to lure private capital into independent power projects (IPPs), the most famous being the hugely controversial Dabhol power project, built near Mumbai, in which the now bankrupt Enron played a key role.

These reformers did not go far enough. Of the two hundred–odd deals signed early in the 1990s, only a handful have come to fruition. Nearly all the projects got tangled in red tape, politics, and severe financial difficulties. Because the IPPs had to sell their power to the bankrupt SEBs, their bankers insisted that the states provide financial guarantees. This was politically and financially difficult given the states' parlous finances. Even those companies that struggled to make IPPs work were lambasted by critics as "robber barons" and "greedy investors who sought to insulate entirely from risk by seeking guarantees and guaranteed returns . . . in a climate of inadequate oversight."

The Karnataka state government forged ahead against the odds. In 2001 it set off on a set of difficult reforms, ranging from expansion of metering to the eventual privatization of the KSEB's distribution business. Other Indian states embarked on similar reforms, responding in part to outside pressures. In the mid-1990s the central government and the World Bank started to shift their focus from grandiose generation projects to the less-glamorous task of fixing SEBs. The World Bank made a controversial decision to offer funds and expertise only to those Indian states undertaking reform of the bankrupt electricity boards.

Predictably, given India's legacy of overregulation and corruption, these reforms got off to a rocky start. The International Energy Agency, which is generally supportive of market reforms, issued a surprisingly critical appraisal of India's efforts in 2002. That report's author spoke with a bluntness usually not found in agency-speak: "How can you expect market reforms to take off in a country where there is no real market in the first place? Every bit of the power value chain—from the mafias at the coal mines all the way through to the pilferage at the distribution system—is riddled with corruption. There is absolutely no accountability whatsoever!"

Navroz Dubash of WRI thinks that India's "state-level reforms have produced mixed efforts at best." What is more, Dubash argues that the road of energy reform looks to be a bumpy one in poor countries generally. A WRI-led study of power-sector reforms in half a dozen developing countries, including India, reached a sobering conclusion:

> In all the case study countries, the shift to a more decentralized, market-led approach has brought no broad vision for future development of the sector . . . By focusing on financial health, reform in the electric power sector has excluded a range of broader concerns also relevant to the public interest . . . socially and environmentally undesirable trajectories can be "locked-in" through technological, institutional, and financial decisions made now that constrain future choices.

The authors advocate a more participatory reform process—as opposed to one rammed through by World Bank edicts or backroom deals done among political elites. However, they admit to the challenge of implementing their own proposal: they accept that "more complex processes bring with them greater risks of capture by special interests and failure due to a cacophony of voices." In other words, no matter which approach to reform you take, it isn't going to be easy.

Bombay Dreams

It has become clearer than ever before that it is the rural rich, not the subsistence farmer too poor to afford a water pump, who benefit from the billions of dollars spent every year on subsidizing power for agriculture. Many of the electricity "losses" attributed to unmetered consumption in agriculture never reach farmers. Commercial and residential users, in cahoots with crooked SEB officials, pilfer that power. The best proof of this once-contentious notion is in Delhi: the local SEB has virtually no agricultural clients, yet investigations into its books a few years ago revealed that it "lost" a staggering 55 percent of its power. The myth of the politically untouchable Indian farmer is finally being debunked, and consumers have started to grow impatient: impatient with power cuts; impatient with power losses; above all, impatient with the failed promises of populists. They seemed to be giving reformers a chance to do better. In the Indian state of Uttar Pradesh, efforts assisted by the World Bank to reform the SEB led to a noisy strike by the state's electricity union in early 2000. Yet the union found no popular support. In elections that followed in Andhra Pradesh state, Chandrababu Naidu, the state's reform-minded chief minister, faced an opponent who promised free electricity. Naidu stuck to his guns: "Get free electricity for one hour a day, or pay and get power twenty-four hours a day." Against all expectations, voters reelected him.

Indian consumers, even poor ones, are willing to pay for electricity, so long as they get enough of it. Experience in the state of Rajasthan demonstrated that farmers are happy to be metered, and to pay higher tariffs, if they see a reliable supply. This is true of the urban poor too, as a visit to Maratha colony, a slum near the airport in Mumbai, made clear. The dwellings are slapdash, the tenants have no proper title to their land, and children play among piles of rotting garbage and open sewers. It hardly seemed a promising market for private power. Yet peering into the tenements, I found that each had electricity: a bulb and a fan, perhaps a small fridge or television. Bombay Suburban Electric Supply (BSES), the private firm that serves Maratha, estimated that perhaps 60 percent of its 1.8 million customers live in such slums.

I was taken aback. I asked R. V. Shahi, then the head of BSES, how he could possibly turn a profit servicing slums. He explained with pride that his firm was unique in India in that it had been providing power privately to Mumbai for more than seven decades. The firm had thus avoided becoming ensnared in red tape or populist quagmires like power giveaways to favored groups. Because BSES was run like a proper corporation, he explained, it had invested heavily to provide even these downtrodden consumers with "safe, reliable, quality power"—and they returned the compliment, he explained, by paying their bills. I remained unconvinced. Then his field managers took me on a detailed tour of the innards of the slum's electrical system. They showed me how every home was metered with tamper-proof equipment of the sophistication you would expect in a rich country. They explained that the bills were computerized and that the quality and quantity of energy delivered were carefully monitored. What about the culture of corruption? My guide grinned broadly as he explained, "No problem. We cross-check the meter readings with surprise audits!"

As I was jotting all this down, a curious resident of the Maratha colony stuck his head out of his doorway to see what all the fuss was about. I seized the opportunity to put the BSES propaganda to the test. I ducked into his humble apartment and asked him if I could see an electricity bill. Within moments he produced an envelope containing a computerized printout detailing his power usage and charges. Ah, I thought to myself, now I'll look him straight in the eye and ask him the zinger: Do you really pay the bill? I was sure that such an obviously poor resident of an urban slum, in a country rife with corruption and pilferage of electricity, must have some misgivings. He was genuinely astonished by the question: "What? Do you think the power will come for free?" Even Bangalore's Internet billionaires could not put the case for market-minded reforms more succinctly.

Hope can be found in the dramatic shift, linked to the move to competitive energy markets, away from big energy projects at international financial institutions and aid donors. It can also be found in the growing power of grassroots activists. Taken together, these changes suggest that the centralized, fossil-fuel-based power grid that is the mainstay of developing economies may be surpassed in the future by a much nimbler and far cleaner energy infrastructure that can serve the needs of the poor. Nowhere is this convergence of forces more apparent than in the heated debate over the future of big dams.

Dam Shame

"Big Dams are to a Nation's 'Development' what Nuclear Bombs are to its Military Arsenal. They are both weapons of mass destruction. They're both weapons governments use to control their own people. Both twentieth-century emblems that . . . represent the severing of the link, not just the link—the *understanding*—between human beings and the planet they live on." So says Arundhati Roy, the celebrated Indian novelist turned protester. Those are controversial—if not explosive—charges she's lobbing at hydroelectric power, and at first blush, they seem pretty absurd.

Jawaharlal Nehru, a hero of India's independence struggle and its first prime minister, felt that dams were "the temples of modern India." A third of the world's countries—including some of the poorest—depend on hydropower for more than half of their electricity. A substantial portion of the world's irrigated land depends on dams too. Reliable and inexpensive irrigation water helps farmers and ensures that consumers benefit through cheaper food.

In fact, this seems to be the era of the Big Dam. Asia is home to much of the world's dam building, including two of its most ambitious projects: those in India's Narmada Valley (which so outraged Roy) and China's Three Gorges. Indian and Chinese officials are quick to point out the many splendid benefits of these megaprojects. The Three Gorges dam is designed both to generate some 20 gigawatts of electricity and to help control the floodwaters of the temperamental Yangtzi River. India's planners promise that the Narmada project will likewise deliver electricity and irrigation. Such ambitions are admirable. Lack of irrigation relegates millions of Indians to subsistence farming. In China, massive floods in 1998 destroyed five million homes.

So is there any reason to take Arundhati Roy seriously? Actually, she's onto something. All across the world, and even in India and China, a backlash against the Big Dam is gaining momentum. Ecological, social, and financial arguments can be made against dams, but they all point to the same conclusion: megapower is going out of fashion. And, while activist-novelists are unlikely to admit it, it is the liberalization of the poor world's energy sector that is dealing the deathblow to these monstrosities.

Large dams were once welcomed by some environmentalists as a source of clean and endlessly renewable energy, but their green credentials are tarnished. The flooding that accompanies big dams in the developing world usually submerges large tropical forests. As vegetation decays, it can release lots of methane, a much more powerful greenhouse gas than carbon dioxide. Unlike giant dams in rich countries, dams in the tropics must endure the ravages of monsoons. One common result is silting up, which may cut the original generating capacity by 70 percent or even 80 percent within a few decades.

Many species are affected by hydroelectric projects, most famously the salmon in America's Columbia River. Big dams can suck nutrients out of rivers, change their temperature, alter their flow, and distort flood cycles—all of which harm some species living in the watershed. People can be hurt too, as some reservoirs act as breeding grounds for mosquitoes and therefore nasty diseases like malaria.

Putative dam builders must also contemplate the potential social cost: as many as 80 million people worldwide have been displaced by dams in recent decades. Governments usually make noises about giving them cash or land in compensation, but in practice, such promises are often forgotten. Given the steady advance of democracy around the world, it's getting hard for central planners to dismiss the voiceless millions. The Narmada project has displaced many tens of thousands. Pressure from grassroots groups has already forced the government to scale back its plans. The partially finished Sardar Sarovar dam, which is at the heart of the scheme, will probably never be completed. Of the several thousand dam-related projects in that valley, most will never see the light of day. Dams that are finished will be at a lower height than envisioned, and they will be redesigned to reduce their social and environmental impact.

Even in China, the big-dam juggernaut is not so formidable as it once seemed. Doris Shen of the International Rivers Network observed that the Chinese government was having terrible difficulties selling the electricity from earlier dam projects whose power proved much more expensive than power from smaller stations. The further China's electricity liberalization goes, the less competitive power from such white-elephant projects will be. Indeed, as the true costs of the Three Gorges dam soar skyward, Shen is convinced that only the hubris of an all-powerful regime—along with some kickbacks—has saved the project.

The economics of big dams make little sense. Even Asia's zealous dam builders "are being pulled into global principles by market forces," according to Achim Steiner. Today he heads the World Conservation Union, but he used to be the secretary-general of the World Commission on Dams, an innovative organization created by governments, development agencies, and NGOs to help forge a consensus on the big-dam problem. That commission's final report made it clear the future for big dams is not bright.

Large hydro-projects are being squeezed by both private and public investors. Aid money is drying up. Big dams are so controversial that even the World Bank, once the biggest force behind them, has grown skittish. The agency suffered a humiliating setback in the 1990s in India when an outside report it commissioned turned out to be sharply critical, leading it to abandon the Narmada project. That in turn prompted a broader internal review, which concluded that governments often fail "the acid test" the World Bank recommends for dams—"the restoration of incomes and standards of living of project-affected people."

Less aid money means higher costs of financing. And the inevitable protests and legal wrangles with NGOs adds financial risk, which also translates into higher costs. The sharpest blow, however, comes from the continuing deregulation of the global power industry. Why? Simply put, that shifts financing away from the state sector, where bureaucrats spend other people's money, to the private sector, where people spend their own money. The predictable and happy result is a move toward lower-risk projects with more reliable, quicker returns. Smaller plants fired by natural gas and even renewable energy are finding favor over dams and nuclear power plants.

Even if big dams are out, small ones need not be. They can achieve many of the benefits promised by big dams at a fraction of the cost—and with the support of locals. So it's not the end of hydropower. It's just one more part of the global move from megapower to micropower.

Small Is Beautiful

Central planners have lavished vast sums on grandiose schemes to build dams and nuclear- or coal-fired plants. Because these projects had little accountability or transparency, they have typically been completed late, over budget, and with a far greater social impact than estimated, and they are usually run at far lower efficiency levels than promised.

Market reforms in the production and delivery of power have injected a strong dose of reality and risk assessment. Most of the financing of power projects in developing countries now comes from the private sector, a dramatic reversal from just a few years ago. As a result, from Honduras to the Philippines, energy investors are increasingly favoring small, efficient power plants fired by natural gas or other forms of distributed generation, bypassing the dinosaurs.

Liberalization is also leading to the reduction of subsidies for fossil fuels, which is sure to boost the penetration of clean forms of distributed generation. While the rich world is also guilty of such perverse distortions of the market (Germany and Spain still subsidize their coal industries, for example), the IEA thinks that the bias in favor of dirty energy sources is greater in many poor countries. A reliable study of eight of the biggest energy-consuming countries outside the rich world found that the average end-use prices for fossil fuels were about 20 percent below what market prices would have been. Removing these subsidies, it estimates, would reduce wasteful energy consumption by 13 percent, increase those countries' economic efficiency significantly, and even lower emissions of carbon dioxide by 16 percent. All told, the IEA experts think the GDP of those countries could be lifted by 1 percent or so—a big enough number to matter.

Piously claiming to defend the interests of society's poorest, governments in developing countries still lavish billions of dollars in blanket subsidies for gasoline, kerosene, and coal, and they waste outrageous amounts of electricity by giving it away free to politically important groups like rich farmers. Rarely do the truly indigent—landless peasants or laborers—benefit. Worse yet, argues the IEA, "the waste and inefficiency that subsidies have engendered in the electrified part of the economy have often blocked the expansion of the electricity grids."

Things are starting to change. Prodded by market reforms as well as by growing concerns about the environmental impact of fossil fuels, governments are removing some of these subsidies. The most dramatic example is China, which has stripped away many of its subsidies for dirty coal and has encouraged a shift to cleaner natural gas. Though reliable figures are hard to come by, coal consumption in China appears to have moderated over the past few years even as its economy has grown by leaps and bounds.

Development banks have traditionally been among the biggest backers of giant power projects in poor countries. Now they are leaving such decisions to the market, pushing governments for deregulation and privatization of their power sectors. Donors are also taking into account the high interest cost incurred by inevitable construction delays, and taking aim at export credits, another hidden subsidy for dirty power projects.

Encouraging as these top-down developments are, the best reason to believe that a cleaner energy future awaits the world's poor comes from the bottom up. Conventional wisdom holds that poor countries will not care about the environment or cleaner energy—and will certainly will not pay for it—as long as the masses are mired in poverty. Tell that to the Grupo de Cien, an environmental group in Mexico City that has been pushing for a cleanup of that city's foul air, or the Centre for Science and Environment (CSE), an environmental NGO in Delhi whose noisy and well-coordinated campaign to improve children's health led to a ban on diesel buses. Local activists all over the developing world, emboldened and enriched by activist brethren in the rich world, and informed and linked by the Internet, are clamoring for greenery.

Perhaps surprisingly, this fervor is spreading to China too. Liang Congjie, the head of Friends of Nature, runs one of China's few NGOs. He explains why he is allowed to operate: the Communist bosses have judged that public outrage over pollution is so strong that they allow complaints about the environment to be voiced. Even the media are allowed to run riot on the issue, and investigative reports and undercover exposés of green cheats have now become the norm. That is not quite full-fledged democracy, but it's surely a step in the right direction.

Village Power: Meet Market Power

The soaring demand for energy in the poor world can be met in ways that empower people rather than bureaucracies, that are cleaner and ultimately more sustainable than today's. These are the foundations of a philosophy that is emerging as the energy paradigm for the next century. At a recent summit in Washington, D.C., dubbed Village Power, experts from donor agencies like the World Bank and America's Agency for International Development met with nongovernmental groups devoted to clean energy, as well as with private-sector companies promoting renewable energy.

The rallying cry for the gathering came from Frank Tugwell, the head of Winrock International, an Arkansas-based development NGO, who called on the attendees to commit to a specific target: "We must provide electricity to 500 million people currently without access to it by 2010 through investment in community-based renewable energy!" That is a noble goal, and certainly a pressing one. Making it reality, however, will be impossible unless the powers that be pay close attention to the lessons offered by the contrasting attitudes of Bill Gates and Gerardo Zepeda.

So who was right? India's Centre for Science and Environment weighed in on this matter recently. It ran a cover story in its magazine called "Computer Connected Villages: Is It Real?" After a detailed investigation of San Ramon–style experiments all over India, the group's researchers came up with this analysis: "No, IT [information technology] will not usher in the dawn of a brave new world, transforming Indian villages into stuff that makes advertising brochures. But yes, IT is a great tool, which, if properly used, could improve the life of millions. But some daunting problems need to be conquered first . . . [and] the greatest worry is electricity supply." The researchers pointed to some examples of makeshift power sources used to make those fancy computers work under Indian conditions: "A project in Pondicherry is combining the power supply from the grid with battery backup and solar power. Draught power from livestock can be used to generate 40,000 megawatts of power in villages." Farm animals are probably not quite the sort of "village power" that the big shots in Washington had in mind, but evidently they do the trick. In other words, the Indian example suggests that Gates and Zepeda have each got it partly right.

Gates was certainly spot-on in pointing out that cost is often an obstacle: Who will pay for the fancy solar panels and other expensive equipment that undoubtedly improves lives? In stressing that many of the world's poor earn but one dollar a day, he made two good points: their primary need is for food and clean water, not snazzy electronics; and even accepting that access to electricity improves their chances of obtaining those necessities, subsidies are clearly required.

Zepeda, for his part, was right to stress that electricity can transform remote villages, helping them leapfrog from the Dark Ages to the twenty-first century. In San Ramón, locals started to take advantage of telemedicine and distance learning. The town's women learned how to peddle their arts and crafts through e-commerce. Some villagers earned money teaching Spanish, via video link, to students in Oklahoma (which had a shortage of Spanish teachers). An educational software package even identified several gifted children who went on to higher education.

However, both men were also wrong in important ways. Gates's pessimism, for example, oversimplified the matter. While the very poorest will always need subsidies, that does not mean there is no ability to pay for energy among the world's poor. Of the people without access to energy today, the World Bank estimates that perhaps half do have the ability to pay commercial rates for electricity. The rest, of course, will clearly need some government subsidies however the power is delivered.

The crucial—but often overlooked—point is that there is ample evidence that the poor already do pay, often heavily, for energy services that the state fails to deliver. The amount of money spent by those dirt-poor households on inefficient, dirty ways of delivering energy—such as kerosene, candle wax, and recharging batteries—can be greater per kilowatt than what is spent by middle-class urban households or wealthy farmers on heavily subsidized grid electricity. Families in Peru's remote highlands now pay some $4 a month for candles. For just a bit more, they could afford the much-higher-quality power offered by village power units: experts say that local entrepreneurs can turn a profit by leasing out a small 35 kW solar unit (enough to power two bulbs and a radio) for just $80 a year.

In Yemen, where there are no legal barriers to entry into the "wires" business, dozens of tiny private generators have sprung up to service households not reached by the inadequate grid system. Though the price charged by these entrepreneurs is quite high and the frenzied competition occasionally unruly, even poor households scrape together the funds rather than live in the dark. As a result, electricity penetration in Yemen tops 50 percent of households, a far higher level than found in comparably poor countries. Yemen's remarkable experience suggests that there may be an enormous market in serving the energy-poor; it also highlights the gross inefficiency and unfairness of centralized grid systems.

While Zepeda's high-tech model is inspiring, some worry that it will simply prove unsustainable. What will San Ramón be like a few years from now, after the spotlight has faded and the government has either lost interest or lost power? Indeed, Zepeda himself is no longer a government minister. Zepeda insisted that the changes in the village are both cost-effective and irreversible, but development experts are more skeptical. "The developing world is just littered with examples of energy projects that have failed because donors or governments did not think about how they will be maintained and paid for," said Christine Eibs Singer of E+Co, a nonprofit organization specializing in financing renewable energy. She insisted that the key to sustainability lies in helping local entrepreneurs create markets for energy services. That is not as far-fetched as it may sound. Her outfit has been at the forefront of an innovative approach that helps locals to help themselves. Using aid money from the Rockefeller Foundation and others, E+Co invests "risk capital" in start-up firms in developing countries that want to deliver clean micropower to villages; it also advises them on how to craft sensible business plans, sales and distribution strategies, and so on.

Bankers in developing countries refuse to accept the fact that the poor are usually excellent credit risks: the dozens of projects funded by innovative NGOs like E+Co have demonstrated repayment rates of 92 to 98 percent. Obviously, users must pay for services. Local firms have employed various approaches, ranging from up-front cash for some customers to fee-for-service or leasing approaches. A particularly promising technique, similar to that used by some mobile-telephone firms, employs prepaid tokens that can be purchased at shops. That eliminates the need for awkward credit checks and large up-front payments.

TERI's Raj Pachauri thinks that helping develop local entrepreneurs with incentives to maintain and expand the micropower infrastructure is the key. His group has worked for years to come up with techniques and tools for training such local micro-capitalists. He tells of the sheer astonishment expressed by an Indian cabinet minister when Pachauri took him to an extremely remote tribal village in Bihar, a particularly poor state in the east of the country: "He had seen the situation in the village earlier, when all the homes were full of soot and smoke. After the renewable energy interventions, he proclaimed that what he had seen was a silent revolution. He immediately directed his ministry to enhance resources for such programs."

Unfortunately, this successful model is not easy to scale up to the level of the Indian federal government—let alone the World Bank or big donors, "600-pound gorillas" that are accustomed to dealing directly with national governments on projects worth tens or hundreds of millions of dollars and are simply unable to disperse small loans to entrepreneurs. Through "investment pools" and other innovative approaches, E+Co is now hoping to reach small energy entrepreneurs (or perhaps even to create them) in poor countries neglected by the capital markets and those big gorillas.

One approach is the "negative concession," through which governments privatize the provision of basic services to the very poor or to scattered rural populations, awarding these contracts to the bidder that requires the smallest subsidy. Chile considered the use of this approach in water provision, and Nepal in telecommunications. South Africa embraced this concept in a promising way too: rather than expand the central power grid to remote areas at a cost of a few hundred dollars per household, the government declared that it would offer that subsidy instead to the firms that win concessions to provide off-grid, renewable power to those villages now without grid access.

So is it time for the world's energy-poor to start cheering? José Goldemberg, a Brazilian expert in energy poverty, offers these words of caution: "We cannot simply ignore the energy needs of the two billion people who have no means of escaping continuing cycles of poverty and deprivation . . . But changing energy systems is no simple matter. It is a complex and long-term process—one that will require major and concerted efforts by governments, businesses, and members of civil society . . . we need to do more to promote energy efficiency and renewables, and to encourage advanced technologies that offer alternatives for clean and safe energy supply and use. We also need to help developing countries find ways to avoid retracing the wasteful and destructive stages that have characterized industrialization in the past."

In sum, we need a happy collision of clean energy, microfinance, and community empowerment. Mobilizing investor interest will be critical to unleashing the power of markets. So too will intelligent government policies, argues WRI's Navroz Dubash: "Market reforms will support a transition to a micropower future only if reforms are intentionally designed to do so." And the good news is that this is beginning to happen. Micropower is beginning to join forces with village power.

Grace Yeneza of Preferred Energy, a Philippine NGO, could not contain her excitement as she described the success her group has had implementing this approach in several remote villages along a small river in the highlands of Luzon. "These villages had no access to the electricity grid," she explained. Working with the local councils called *barangays*, her group built a micro-hydroelectric plant that delivers electricity to the common areas of several villages through a mini-grid. Donor agencies paid for the equipment, but she said the key was the fact that the villagers pitched in "equity" in the form of labor and local materials. They also organized themselves into a management committee to run the plant. Those who want power for their households can receive it, but they must pay for the privilege. Then she got to the best part of her story: "Thanks to this project, the villages not only got power but they also learned to cooperate and stop fighting over the river!"

With more success stories like that, even Bill Gates and Gerardo Zepeda would probably agree that the future for the world's poor need not be so dark after all. Or to put it in Arundhati Roy's words, "The dismantling of the Big. Big bombs, big dams, big ideologies, big contradictions, big countries, big wars, big heroes, big mistakes. Perhaps it will be the Century of the Small." And how fitting it would be if the fuel of choice for that century were to be the lightest element of all—hydrogen.

Epilogue: The Future's a Gas

AS FAR BACK as 1874, during the foulest days of the industrial revolution, Jules Verne envisioned a world based not on fossil fuels but on clean hydrogen energy. He wrote in *The Mysterious Island*:

> Yes, my friends, I believe that water will one day be employed as fuel, that hydrogen and oxygen which constitute it, used singly or together, will furnish an inexhaustible source of heat and light, of an intensity of which coal is not capable . . . water will be the coal of the future.

Could the man who forecast the development of such technological marvels as submarines, helicopters, and space travel have gotten energy right too?

Fuel cells are as good a bet as any to deliver that energy, given the dramatic progress that has already been made in bringing them to market, but the beauty of the hydrogen model is that it is not wedded to any specific primary energy source or technology.

Hydrogen can be made from fossil fuels as well as from renewables; and it can be used in internal combustion engines as well as in fuel cells. That flexibility, as much as its cleanliness, ensures hydrogen a bright future.

Of course, no one knows what the future holds for our planet. A century from now the world may continue to be happily, if unhealthily, addicted to fossil fuels. Still, after a century in which talk of hydrogen energy has been strictly the domain of crackpots, there is at last conclusive evidence to think that Jules Verne was onto something. It turns out that the world has slowly been shifting away from heavy hydrocarbons like coal and toward lighter ones like natural gas for many years. In molecular terms, all are combinations of carbon and hydrogen—but today's fuels have less carbon content and so burn cleaner. Hydrogen, which has no carbon attached and can produce electricity without any emissions whatsoever, is the ultimate energy carrier.

Jesse Ausubel of Rockefeller University describes the trend this way: "The most important, surprising and happy fact to emerge from energy studies is that for the last 200 years, the world has progressively favored hydrogen atoms over carbon . . . the trend toward 'decarbonization' is at the heart of understanding the evolution of the energy system." Even before industrialization, societies started shifting from dirty solid fuels with a high carbon content to liquid hydrocarbons and ultimately to clean-burning gases, shifting from wood and cow dung to coal to oil and natural gas. Comfort, convenience, and cleanliness have long driven the decarbonization trend—and these are the best reasons to think that the world will one day reach the hydrogen era.

It is encouraging that the most powerful force for decarbonization through the ages has been the free will of ordinary people, expressed through choices made in the free market. It is no coincidence that the historical trend has stalled in recent decades, when governments have taken to meddling heavily in energy markets. Robert Hefner of the GHK Company, an American energy firm, observed: "For more than a hundred years, free markets and the ingenuity of mankind worked efficiently to decarbonize our energy systems. It was only in the 1950s, when governments began to tinker with price controls and later, reacting to cries of shortages by the energy industry, allocated fuels among sectors of consumers, that we began to recarbonize the energy system." He is not exaggerating. Many governments, including America's, actually banned the use by factories of natural gas in the 1970s, mistakenly believing that the clean and abundant fuel was scarce. That needlessly led many companies to rely on dirtier power from coal.

If governments level the energy playing field, the trend away from carbon could well resume. Hydrogen thinkers like Hefner argue that by 2050, natural gas and hydrogen will surpass oil and coal as the energy carriers of choice, and that by the end of this century, these energy gases could have 75 percent of the global energy market, the same as King Coal in his heyday. That vision is increasingly finding converts even among mainstream forecasters at government agencies and companies.

The International Energy Agency argues that the great hope for environmentally benign energy lies in "crosscutting" technologies that can exploit energy from several different sources. In particular, the IEA highlighted hydrogen fuel cells and carbon sequestration as technologies that, alone or together, could "have a profound impact on the long-term prospects for energy supply." Royal Dutch/Shell, a world leader in scenario planning, was even more enthusiastic about the prospects for a hydrogen economy. In late 2001 the oil giant unveiled two energy scenarios that it reckons could unfold by 2050. One sees fossil fuels and incumbent technologies maintaining supremacy over time, but the other predicts a dramatic burst of innovation and experimentation that could usher in the hydrogen age. According to this second scenario, half of the new cars sold in the rich world in 2025 will use fuel cells, and oil demand will slacken to the point that the commodity becomes dirt cheap. OPEC: you have been warned.

The New Geopolitics of Energy

The Shell planners provocatively suggest that China and India could decide to embrace the hydrogen economy wholesale as a response to environmental concerns and growing insecurity about imports from OPEC. Those countries could tap their vast stores of domestic coal to unleash the energy trapped inside, sequester the carbon, and use the hydrogen to power fuel-cell-driven cars and electricity plants. Is it ridiculous to think that poor countries, where many people still use dirty solid fuels such as wood and cow dung, might leapfrog to the hydrogen age? China's leaders, concerned about the country's growing reliance on imported oil and gas and eyeing the opportunity to develop a homegrown fuel-cell car industry, are now pouring money into hydrogen research. That is encouraging, but of course it is unreasonable to expect poor countries to pay for pricey new technology on their own; the rich world must lead the way.

If the giants of the developing world really embrace clean micropower, the impact could be dramatic. Such a move would not only ease worries about local pollution and global warming but also dramatically alter the geopolitics of oil regions like the Middle East. How much military interest will America really have in Saudi Arabia, say, in a world in which oil, whose reserves are highly concentrated, is replaced by hydrogen, which can be produced in a highly distributed fashion?

We have a chance to set the energy system on a more sustainable footing. However, that chance could slip away forever if the Chinas and Indias of the world build thousands of old-fashioned coal plants—without gasification, sequestration, or any other clean technologies—that would lock them into dirty energy and local air pollution for fifty years or more, and would cast a grim shadow over the coming century of global warming.

The good news is that, after decades of stagnation, the energy realm is wide open to new ideas, new technologies, and especially *new ways of thinking*. Or, as in the case of the Sage of Snowmass, insights that are simply ahead of their time.

Imagining a Brighter Future

The view from the ivory tower is breathtaking. I know, because I worked at *The Economist*'s London headquarters for many years. Whenever a big shot comes over for an interview, we journalists try to book a meeting room on the very top floor of the architecturally distinctive cream-colored tower. The unobstructed views of Big Ben, London's West End, and the Thames River never fail to impress even the heads of state who occasionally stop by.

So I was not at all surprised when Amory Lovins went straight to the window when I brought him up to the St. James's Room one sunny afternoon. The energy guru seemed transfixed by Buckingham Palace, which is just a stone's throw from our building. So, I chuckled to myself, even Amory Lovins gets starstruck! As I looked more carefully, though, I realized that he was actually engrossed by the rooftop of the squat modern building next door, with its exposed ventilation shafts, cooling equipment, and other bits of energy infrastructure. And he was not impressed: "Just look at those fans—half are running flat out, and half are off. The energy used rises with the cube of the air speed, so they should really run them all at half speed instead and save three-fourths of the electricity. Aargh! Are those sharp bends on those ducts? And just look at those tall and skinny cooling towers. They should be using short and fat ones instead." He was oblivious to the architectural glories of the house of Windsor.

Lovins deserves his reputation as a man who looks at the world a little differently. "I guess I wear what the Japanese call Muda spectacles," he says with a twinkle. "They find purposelessness, waste, and futility." Three decades ago he made a name for himself by suggesting—quite controversially—that energy use need not grow in lockstep with economic output. If energy efficiency was encouraged, he argued, economies could follow a "soft path" instead. Along with Walt Patterson of Britain's Royal Institute of International Affairs, another energy thinker whose predictions were originally ridiculed but later proved correct, Lovins has long argued that what ordinary people want are energy services like cold beers and hot showers, not energy for its own sake. Once the goal of policy is defined that way, it becomes clear that meeting people's needs by saving fuel can often be smarter than building ever more power plants, pumping out ever more oil, or cranking out ever more electricity. Or as Lovins likes to put it, " 'Negawatts' are often cheaper than megawatts." And, of course, he has argued for years that the expensive and inflexible gigantism of the power industry must yield to micropower—and he has shown how that trend will upend the auto industry by confronting it with hydrogen-fuel-cell technologies like his beloved Hypercar. Astonishingly, the energy world, which has been stuck in a monopoly mind-set for ages, is at last beginning to listen to such radical ideas. For the first time in decades, there is reason for hope about the world's energy and environmental future. Since most environmental problems can be tackled if enough clean energy is available at a reasonable cost, fixing energy is really the key to sustainable development.

Lovins's most important contribution to the energy debate is his faith in mankind, which contrasts sharply with the gloom and doom peddled by most greens. In fact, though he clearly shares the aims of most environmentalists, he positively rankles at being called one: "Actually, I prefer elegant frugality, not a hair shirt. I don't mind taking an infinitely long shower using solar energy, because it doesn't waste anything. I say live well—but take nothing, waste nothing, and do no harm . . . In an odd way, I agree with neo-Cornucopians that the key to the future lies in innovation and ingenuity." That's a vision for the future of the planet that we all can share.

However, his optimism will be warranted only if we have the courage to tear down the many market distortions—ranging from unfair coal subsidies, to monopoly protection for stodgy incumbent utilities, to regulatory barriers impeding micropower and hydrogen—that stifle innovation and ingenuity in the energy realm. Getting energy prices right, and sweeping away the obstacles that make it hard for ordinary people to respond to those price signals, would be the single best way to nudge the world toward a brighter future. But old ways of thinking die hard, as America's experience with energy taxes suggests.

Taxing Times Ahead

American environmentalists are big fans of legislation that prods car companies to produce fewer gas-guzzlers. Though tighter rules clearly will have some impact, merely tinkering with fuel-efficiency standards is unlikely to do much to push the world toward a clean hydrogen economy—or any other as yet unforseen technological future independent of oil. Direct environmental taxation is a cheaper (assuming those revenues are "recycled" to taxpayers) and more effective way to level the energy playing field than such mandates.

Much of Europe recognizes this and has over the past decade started to shift the burden of taxation from income to environmental aims such as reducing carbon emissions. Tax remains a four-letter word in the United States, but Arizona could yet point the way. To encourage greener cars, the Arizona legislature recently offered a tax credit to citizens who bought a "dual fuel" car capable of using either gasoline or natural gas, which burns cleaner. The offer proved so wildly popular that it cost the state some $500 million, not the $5 million budgeted, before officials could shut it down. Though this scheme was flawed in several ways, its success makes abundantly clear that even car-loving Americans are willing to embrace greener technology if government sends the proper price signal. And the best way to do that is through taxation that reflects all the costs—in terms of national security, as well as those costs to the environment and to human health—of burning oil.

Unless America tackles its petro-addiction, there is little hope that the rest of the world will get off oil. So what are the chances that America will change its tax policy? James Schlesinger, an old Washington hand who has served as both Secretary of Energy and Secretary of Defense, offers a lesson from the past: "An energy tax is surely the best approach—but I've still got the black-and-blue marks from the times I have proposed similar measures in the past! When energy prices are falling, it becomes very difficult to persuade the public to accept long-term measures."

One recent fiery encounter on Capitol Hill suggests that times have not changed much. In 2001 a panel of experts from the National Academy of Sciences issued a report that undermined the auto industry's insinuation that tightening efficiency standards would prove the end of the American way of life. The panel's chairman, Paul Portney, was asked to address a congressional panel. George Allen, a senator from Virginia, was quite clearly unhappy with the report's suggestion that fossil-fuel use could be easily curbed by tightening regulations. Portney, who is the head of Resources for the Future, a nonpartisan think tank in Washington, was asked whether there is any other way to encourage fuel efficiency without resorting to market-distorting regulations.

Much to Allen's surprise, Portney responded, "Why, yes." What was his magic solution, wondered the senator. The swift response: "A significant increase in the federal gasoline tax." That sensible reply so outraged Allen that he could only manage to grunt that the notion was "just flat ignorant." Senator John Kerry, the panel's chairman, was clearly embarrassed at the attack on the distinguished visitor. Making a pointed reference to the fact that Americans pay less than a quarter the amount of gasoline tax paid by many across the Atlantic, Kerry said in exasperation, "I can see the headlines tomorrow: 'Virginia Senator Calls Europeans Ignorant.' Or maybe worse . . ." Yet Allen remained unrepentant.

Still, Big Oil is swimming against the tide of history. Political leaders with foresight would see that genuine energy independence will come not from adding a spoonful of Alaskan oil or a dollop of conservation (never mind just leaving things to the distorted workings of the oil market), but from encouraging the speedy development of alternatives to petroleum. And powerful signals—such as carbon taxes, which tell the market that the environment matters—must be an important part of the strategy.

Hydrogen energy will not arrive overnight, but it has the potential to do many good things. If cars were powered by fuel cells using clean hydrogen, noxious emissions and greenhouse gases would decline sharply. As small heat-and-power units spread to homes and offices, the inefficiency of dirty coal-fired power plants would be exposed. Poor countries have the most to gain from this efficient, flexible, and—eventually—cheap technology. As ordinary people everywhere see these benefits, they may even be able to forget about the H-bomb and the *Hindenburg*.

Not-So-Mysterious Islands

Looking across Reykjavik's waterfront at 3 a.m. in July, you might expect to see a bright sky lit by the Arctic sun. Yet what you are more likely to encounter is a shroud of unhealthful, unattractive smog hanging over the otherwise spectacular vista. It may seem surprising that this land of pure water, cool glaciers, and invigorating hot springs suffers from air pollution, for the country gets nearly all of its heat and electricity from superclean geothermal and hydroelectric sources. However, the problem is transport: all its cars and buses, as well as its huge fishing fleet, still rely on dirty gasoline and diesel. The nasty emissions from those vehicles ensure that Icelanders do not escape the burning eyes and scratchy throats that bedevil urbanites elsewhere in the world.

Bragi Arnason, a wizened academic known to his countrymen as Professor Hydrogen, has been trying to change that for decades. During the oil shocks of the 1970s Arnason argued that his countrymen should produce fuel by using cheap and clean local supplies of electricity to make hydrogen from water—just as it was already doing in order to make fertilizer. At the time, he was derided as an eccentric with his head in the clouds, or even "stupid," as he recalls. His ideas got nowhere for many years, when oil seemed cheap and hydrogen technology hopelessly expensive. The revival of interest in fuel cells globally in the 1990s—and the OPEC oil cartel's revival from the brink of collapse at the end of that decade—suddenly got everybody's attention.

The long-suffering Arnason finally got his way in 1999, when the country pledged to become the world's first hydrogen-powered economy. This attracted Shell, DaimlerChrysler, and Norsk Hydro, a Norwegian energy firm experienced in making hydrogen. With local partners, they have set up a joint venture in the country. The project's first phase introduces Daimler's fuel-cell buses, powered by hydrogen made using renewable energy. After that, Iceland plans to start replacing all its cars and buses, as well as its fishing fleet, with fuel cells powered by hydrogen or methanol. Ultimately, Iceland sees itself exporting both hydrogen and its fuel-cell expertise: "We'll be the Saudi Arabia of the hydrogen economy!" exclaims one local enthusiast.

The Icelanders are not alone. Halfway around the world, the leaders of two other island states are also actively promoting a shift to hydrogen. In Vanuatu, a particularly low-lying island chain, elders are frightened that the rising seas that accompany global warming will mean their homes, habitat, and heritage will be wiped out within their lifetimes. In 2000, the island's leaders asked for help from the international community to transform its energy economy so that by 2020 it replaces its costly petroleum imports with hydrogen made from solar energy. In Hawaii, too, similar moves are afoot to forge a hydrogen economy. The University of Hawaii is actively researching fuel cells, while Hawaiian legislators at the state and federal level have for years been promoting energy bills friendly to hydrogen.

Why are these island states embarking upon this curious experiment? There are several reasons—from a desire to end their reliance on expensive imported petroleum to the need to tackle local air and water pollution. The most idealistic motive is wanting to do their part to combat climate change—which holds the potential, however slim, to wipe some of these islands out of existence.

Even if self-interest has led them to embrace hydrogen energy, there is also a chance that these islanders are—like the proverbial canary in the coal mine—sounding an alarm that the rest of the world would be wise to heed. Climate change will affect every part of the world, whether for good or for ill, and all of us would do well to think hard about how best to adapt.

Stopping the use of fossil fuels completely and immediately would be foolish and needlessly expensive, but a thoughtful, phased shift to hydrogen-fired micropower would not. On the contrary, the innovative technologies unleashed by market liberalization and environmental demands hold out the promise of an inexpensive, and maybe even profitable, transition to a cleaner energy world. If we grasp that opportunity, then there is every reason for hope about our planet's future. Indeed, there is every reason to think that today's nascent energy revolution will truly deliver power to the people.

NOTES

BIBLIOGRAPHY

ACKNOWLEDGMENTS

INDEX

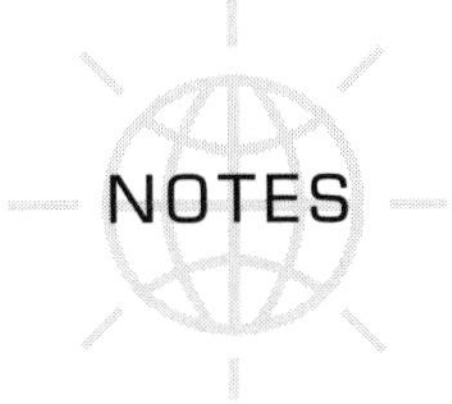

NOTES

Introduction: The Coming Energy Revolution

4 "If you": See *Weekly Petroleum Argus*, October 8, 2001.

7 "Perhaps unsurprisingly": Ali Naimi delivered his speech, "The Geopolitics of Energy and Saudi Oil Policy," at a conference organized by the Center for Strategic and International Studies in Washington, D.C., December 1999.

8 "On one estimate": Donald Losman, "Economic Security: A National Security Folly?" Cato Institute Policy Analysis Number 409, August 2001.

9 "Explicitly citing": "National Energy Policy: Report of the National Energy Policy Development Group," The White House, May 2001.

9 "Even assuming": The International Energy Agency's *World Energy Outlook* series has some eye-opening forecasts of oil trends to 2020.

10 "Still, many": Evar Nering, "The Mirage of a Growing Fuel Supply," *The New York Times*, June 4, 2001.

11 "One efficiency": See the excellent website of America's Energy Information Administration (www.eia.doe.gov).

12 "The car industry": "Effectiveness and Impact of Corporate Average Fuel Economy (CAFE) Standards," National Academy of Sciences, 2002.

13 "Amory Lovins": Amory B. Lovins, "Energy Strategy: The Road Not Taken?" *Foreign Affairs*, October 1976.

16 "That is": Bill Ford's extraordinary forecast of the demise of the internal combustion engine was made at the Greenpeace Business Conference in London on October 5, 2000.

19 "One is": The $2 trillion figure is a rough-and-ready estimate of the turnover of the global energy business by Booz, Allen & Hamilton.

1: Micropower—Thomas Edison's Dream Revived

Klaus Lackner and Walt Patterson were very helpful in preparing this chapter.

24 "One fateful": For more on the encounter between Tesla and Edison, as well as other historical details from that era, see two excellent books by Margaret Cheney: *Tesla: Man out of Time* and *Tesla: Master of Lightning*.

26 "America's National": The National Academy of Engineering's ranking of the century's top technical achievements can be found at www.greatachievements.org.

27 "Even so": The figure of two billion has long been cited as the number of people in the world without any access to modern energy. In the run-up to the Johannesburg Earth Summit in 2002, the International Energy Agency completed the most comprehensive analysis to date on the matter, and it concluded that a more accurate figure is around 1.6 billion. The IEA's *World Energy Outlook 2002* has further details.

29 "On one hand": For short and sharp accounts of excesses of the Insull era, see the Smithsonian Institution's "Powering a Generation of Change" Web-based project, and Daniel Yergin and Joseph Stanislaw, *The Commanding Heights: The Battle for the World Economy*.

31 "One of": The Castens, authors of "Transforming Electricity," have founded various advocacy groups to promote distributed power and cogeneration worldwide. See, for example, the World Alliance for Decentralized Energy.

33 "The biggest": To see a cogent case made for micropower, look to Amory B. Lovins et al., *Small Is Profitable: The Hidden Economic Benefits of Making Electrical Resources the Right Size*.

35 "Sound far-fetched?": A more guardedly optimistic view of micropower comes from the International Energy Agency's *Distributed Generation in Liberalised Electricity Markets*.

36 "The architect": For more on the project by Douglas Durst and Robert Fox, see "Power Struggle," *New York* magazine, June 18, 2001.

38 "Despite that bump": Nth Power, a venture-capital fund devoted to clean energy, produces an annual report about investment in energy technology. This San Francisco–based group was one of the first to spot the similarity between investment trends in this area and those seen in telecommunications three decades ago.

41 "The proliferation": On real-time pricing, see various works by Ahmad Faruqui. A good start is his article with Stephen S. George, "The Value of Dynamic Pricing in Mass Markets," Elsevier Science, July 2002.

42 "Even if": Steve Taub at Cambridge Energy Research Associates has produced a number of thought-provoking reports on the potential and pitfalls of micropower. See "Distributed Energy Resources: A Disruptive or Sustaining Technology?"

44 "Whether run": ABB, despite suffering from financial troubles largely unrelated to its bet on micropower, still produced some farsighted reports on the future of distributed generation, virtual utilities, and the Energy Internet. Search on the company's website (www.abb.com) for "Alternative Energy Solutions" for more details.

2: Enron vs. Exxon—or, The Sleeping Giants Awaken

Daniel Yergin and Robin West were especially helpful in preparing this chapter.

48 "In its heyday": Ken Lay spoke at a conference organized by CWC Associates at Le Meridien hotel in London in mid-1999.

50 "EnronOnline": See "EnronOnline: Louise Kitchen, Intrapreneur," a Harvard Business School case study by Christopher Bartlett and Andrew N. McLean.

59 "The price": Douglas T. Terreson and David Donnelly of Morgan Stanley Dean Witter produced "The Era of the Super-Major" on February 13, 1998.

60 "*The Financial Times*": See the survey of the Energy & Utility Business, *Financial Times*, August 8, 2001, for an analysis of convergence as well as an explanation of why dual gas/power consumers are more valuable than those who sign up for just one utility service.

3: Why California Went B.A.N.A.N.A.s

Dallas Burtraw provided very useful insights in developing this chapter.

67 "By mid-2001": Several columns by Paul Krugman relevant to electricity deregulation appeared in *The New York Times* in 2001: "Enron Goes Overboard," August 17: "Another Useful Crisis," November 11; and "Laissez Not Fair," December 11.

67 "Politicians like": The *Washington Post* ran a series of articles on power deregulation in America. Particularly interesting was "On California Stage, A Cautionary Tale" by Steven Pearlstein, August 21, 2001.

70 "That simple": The IEA's book on power deregulation in the OECD, *Competition in Electricity Markets*, is a good primer on the topic.

71 "Britain's liberal": Alex Benady, "The Green House Effect," weekend magazine of London *Guardian*, March 2, 2002.

73 "In mid-2002": Carl Blumstein, L. S. Friedman, and R. J. Green, "The History of Electricity Restructuring in California," Center for the Study of Energy Markets, University of California Energy Institute, August 2002.

73 "Steve Peace": For more on the "Peace Death March," see the coverage of the state's power crisis in the *Los Angeles Times* (for example, the term appeared in that paper on December 9, 2000, in an article by Nancy Vogel).

75 "California officials": Timothy Brennan, "Drawing Lessons from the California Power Crisis," *Resources*, Summer 2001, published by Resources for the Future.

85 "Nudging down": The study on real-time pricing was released by Ahmad Faruqui on December 18, 2000. Details are available on the EPRI website (www.epri.com).

85 "Severin Borenstein": Severin Borenstein, Michael Jaske, and Arthur Rosenfeld, "A Vision for Dynamic Pricing," in *Dynamic Pricing, Advanced Metering, and Demand Response in Electricity Markets* (The Hewlett Foundation, October 2002).

86 "Even George Bush": See *Things That Go Blip in the Night: Standby Power and How to Limit It* (IEA, 2001).

87 "In the words": The Council on Foreign Relations and the Baker Institute at Rice University unveiled their report, "Strategic Energy Policy: Challenges for the 21st Century," in April 2001.

89 "The best": The figures for National Grid investment are drawn from company sources and estimates by the author.

91 "Mayor Richard Daley": For more on plans to put windmills on top of Chicago's skyscrapers, see a report (on the website of National Public Radio, www.npr.org) by Robbie Harris of Chicago Public Radio that ran on August 26, 2001.

4: Oil—The Most Dangerous Addiction

Morris Adelman, Phil Verleger, and Michael Lynch were especially helpful in preparing this chapter. Books relevant to the entire chapter include Verleger's Adjusting to Volatile Energy Prices *and the IEA's* World Energy Outlook *series.*

95 "For Americans": The pre-invasion debate was captured well in Thomas Friedman, "$6 OR $60," *The New York Times*, July 31, 2002.

96 "In other words": On the prospects for the Iraqi oil sector post-Saddam, see two reports issued by American think tanks: the Council on Foreign Relations and the Baker Institute at Rice University

prepared "Guiding Principles for U.S. Post-Conflict Policy in Iraq," while the Center for Strategic and International Studies came up with "Iraqi Oil . . . The Morning After." Both were released in January 2003.

98 "The most": Jimmy Carter's National Energy Plan was issued by his Administration on April 18, 1977.

99 "What is more": See Michael Lynch, "Oil Market Structure and Oil Market Behavior," WEFA, 2000.

100 "Indeed, Andrew Oswald": See Andrew Oswald, "Unemployment Equilibria and Input Prices: Theory and Evidence from the United States," *The Review of Economics and Statistics*, November 1998.

100 "Philip Verleger": See Philip Verleger, *Adjusting to Volatile Energy Prices*.

103 "There is": For a good snapshot of the international oil industry, see the annual analysis of the world's top energy companies prepared by PFC Energy and by *Petroleum Intelligence Weekly*.

107 "If you": The London-based Oil Depletion Analysis Centre is found at www.odac-info.org. Princeton University's Kenneth Deffeyes has outlined his views in *Hubbert's Peak*.

107 "Their experts": The website of the U.S. Geological Survey (www.pubs.usgs.gov) has the agency's views on oil depletion. The IEA's oil forecasts are built into its annual *World Energy Outlook* forecasts.

113 "Even if": The concern about growing reliance on Saudi Arabia, described in detail in the introduction, can be investigated further by reading the fine print in forecasts such as those put out by the IEA and the EIA. Nearly all such forecasters assume that oil output from Middle Eastern OPEC countries will rise dramatically by 2020 from their levels in 2000. Usually, this forecasting is done by projecting plausible growth in demand and in non-OPEC supply, and by assuming that the shortfall will—hopefully—be supplied by Saudi Arabia and its neighbors.

5: Welcome to Global Weirding

Tom Wigley, James Edmonds, Eileen Claussen, David Victor, Jonathan Pershing, and Michael Grubb were exceptionally helpful in preparing this chapter. Publications that inform the entire chapter include the Third Assessment Report *produced by the UN's Intergovernmental Panel on Climate Change (IPCC), as well as* U.S. Policy on Climate Change: What Next? *by the Aspen Institute.*

118 "Maumoon": Maumoon Abdul Gayoom's views are described in *The Maldives: A Nation in Peril.*

123 "Richard Lindzen": Richard Lindzen's opinion piece appeared in *The Wall Street Journal*, June 11, 2001.

125 "The headline": "Climate Change—It's Faster than You Think," *The Ecologist* special supplement, November 2001.

125 "But that day": Tom Wigley's analysis is drawn from his contribution to *U.S. Policy on Climate Change: What Next?*

128 "America's official": America's National Oceanic and Atmospheric Administration (www.noaa.gov) estimates that three of the last five years of the 1990s were among the "warmest on record."

128 "The group": See "Reconciling Observations of Global Temperature Change" by America's National Research Council, January 2000.

128 "John McNeill": See J. R. McNeill, *Something New Under the Sun: An Environmental History of the Twentieth-Century World.*

129 "Svante": For more on Svante Arrhenius, see William Stevens, *The Change in the Weather: People, Weather and the Science of the Climate.*

130 "*Newsweek* had": The *Newsweek* cover on "global cooling" ran in the issue of April 28, 1975.

133 "That explains": In June 2001 the National Academy of Sciences presented "Climate Change Science: An Analysis of Some Key Questions" to George Bush.

134 "Bill McKibben": Bill McKibben, *The New York Review of Books*, July 5, 2001.

135 "That alarming snippet": The *New Yorker* cartoon by Michael Crawford appeared in the issue of August 5, 2002.

136 "The IPCC": See the *Third Assessment Report* of the UN Intergovernmental Panel on Climate Change.

137 "A few extreme": The Greening Earth Society's Q&A can be found on its website: www.greeningearthsociety.org/faqs_climate.html.

138 "A detailed study": See "Assessment of the Potential Effects and Adaptations for Climate Change in Europe," by the Jackson Environment Institute at the University of East Anglia, Britain.

139 "The biggest impact": For more details on the concerns surrounding ice melt, see the technical summary of the IPCC's *Third Assessment Report.*

141 "Some experts": Thomas Schelling compared the Kyoto process to negotiations on arms control in an article in the May/June 2002 issue of *Foreign Affairs.*

142 "So screamed": *The Independent* newspaper article fingering George Bush as "polluter of the free world" ran on March 30, 2001.

146 "David Victor": See David Victor, *The Collapse of the Kyoto Protocol and the Struggle to Slow Global Warming*, and Scott Barrett's *Environment and Statecraft: The Strategy of Environmental Treaty-Making.*

147 "Not so fast": Michael Grubb's "Keeping Kyoto" appeared in the July 2001 issue of *Climate Strategies* (www.climate-strategies.org).

149 "David Gardiner": The references to David Gardiner and John Weyant are drawn from the Aspen Institute study on climate change, *U.S. Policy on Climate Change: What Next?*

6: Clearing the Air

Arthur Winer, Luisa and Mario Molina, Zmarak Shalizi, Jonathan Pershing, and Richard Morgenstern all deserve thanks for help with this chapter.

161 "The question": The quotes from children are drawn from the California Science Center's website: www.casciencectr.org.

162 "So begins": See *The Southland's War on Smog.*

163 "For centuries": For a scientifically literate yet very readable account of air pollution, see Devra Davis, *When Smoke Ran Like Water*. Her research suggests that the infamous London fog episode of 1952 killed many thousands more than previously believed.

164 "The youngsters": For more information on air pollution in developing countries, see the World Health Organization's website (www.who.int/peh).

164 "Partha": "Are We Consuming Enough?" by Partha Dasgupta, was published by Britain's Economic & Social Research Council in 2001.

164 "Tragically": For more on the link between indoor air pollution and health, see various reports by the environment staff at the World Bank (for example, "Health and Environment, 2001").

166 "Robert Solow": The reference to Robert Solow is drawn from "Sustainability: An Economist's Perspective," published in *Economics of the Environment.*

168 "The British nongovernmental": "Rigged Rules and Double Standards" was published by Oxfam International in mid-2002.

168 "Those words": The quote from the World Trade Organization is drawn from *Special Studies 4: Trade and Environment* by Hakan Nordstrom and Scott Vaughn (1999).

168 "In fact,": The quotations are drawn from a letter to the author from Thilo Bode. They are used with his permission.

169 "Such voices": For more on "eco-footprints," see the WWF's annual "Living Planet" report (www.panda.org) and a critique of that approach by the Danish Environmental Assessment Institute (www.imv.dk).

172 "Such thinking": See *China: Air, Land and Water—Environmental Priorities for a New Millennium*, World Bank, 2001.

175 "The key": See "Demonstrating Emissions Trading in Taiyuan, China," by Richard Morgenstern and others at Resources for the Future.

181 "That is why": See Howard Gruenspecht, "Zero-Emission Vehicles: A Dirty Little Secret," published in RFF's *Resources*, Winter 2001.

183 "For most": Good histories of oil are found in Daniel Yergin's *The Prize* and Stephen Howarth's *A Century in Oil.*

187 "His point": See two publications by Jesse Ausubel: "Will the Rest of the World Live Like America?," Rockefeller University, 1998, and "Resources and Environment in the 21st Century: Seeing Past the Phantoms," Rockefeller University, 1998.

7: Adam Smith Meets Rachel Carson

Robert Stavins, Jonathan Pershing, Daniel Esty, and Jeff Smoller provided invaluable assistance with this chapter. The works by Stavins and by experts at the OECD listed in the bibliography provide an excellent introduction to market-based environmentalism.

198 "A study": The study by Daniel Esty and company can be found at www.yale.edu/envirocenter.

199 "The unlikely allies": That extraordinary opinion piece by Carl Pope and Ed Crane ran in the *Washington Post* on July 30, 2002.

201 "Pinchot to Pinochet": The quote from Gifford Pinchot is drawn from "Green America," which was published by the American Forest Institute in 1980.

203 "Two early": The Wisconsin passage greatly benefited from conversations with Jeff Smoller of that state's Department of Natural Resources.

204 "The most": See *Superfund's Future* by Resources for the Future.

205 "Thilo": Thilo Bode's quote is drawn, with permission, from his letter to the author.

208 "Even so": See "Clean Profits: Using Economic Incentives to Protect the Environment," by Robert Stavins, in the Spring 1989 edition of *Policy Review*.

209 "The title": See "Harnessing Market Forces to Protect the Environment," *Environment*, January/February 1989.

211 "Experts": See various studies by RFF (at www.rff.org) for analyses of the economics of SO_2 trading. Interestingly, part of the credit for declining SO_2 emissions must go to America's deregulation of interstate shipping. One unanticipated result of that market reform was that it became cheap to move coal from the West—which happens to be much lower in sulfur than the Eastern variety—to power plants everywhere.

211 "Sweden's experience": See *Environmentally Related Taxes in OECD Countries: Issues and Strategies* and *OECD Environmental Outlook* (OECD, 2001).

214 "The biggest": See the World Bank's *Greening Industry* report, 1999.

219 "Most impressive": Lord Browne gave a speech, "Beyond Petroleum: Business and the Environment in the 21st Century," at Stanford University on March 11, 2002, that described his firm's progress on its climate-change goals.

8: The Future of Fuel Cells

Special thanks go to Joan Ogden, Robert Socolow, Robert Williams, and Klaus Lackner.

227 "Amory Lovins": See "Hypercars: Materials, Manufacturing, and Policy Implications" and other Hypercar publications from the Rocky Mountain Institute found at www.rmi.org.

228 "Senator": The quote from Senator Tom Harkin is drawn from Peter Hoffmann's book *Tomorrow's Energy*.

229 "The cost": See studies by Arthur D. Little and Andersen Consulting (now known as Accenture).

231 "The proliferation": Figures are from the OECD.

235 "Robert Lifton": See proceedings from the annual Small Fuel Cells conference organized by the Knowledge Foundation.

236 "In early 2002": The reference to the battery bunny is found in a press release at www.jpl.nasa.gov/releases/2002/release_2002_94.html. But is that annoying rabbit powered by Energizer or Duracell batteries? The answer, it turns out, depends on which side of the Atlantic you live on: both companies have used the bunny! The *Economist* letters page (August 26, 2000) broke this important story.

236 "That hurdle": Lance Ealey and Glenn Mercer, "Tomorrow's Cars, Today's Engines," *The McKinsey Quarterly*, 2002, Number 3.

237 "Without doubt": The NECAR story appeared in *The Hydrogen & Fuel Cell Letter*, July 2002.

239 "Even if": See "Hybrid Autos Quick to Pass Curiosity Stage," *The New York Times*, January 28, 2003.

240 "Robert Williams": Joan Ogden, Robert Williams, and Eric Larson, "Toward a Hydrogen-Based Transportation System," www.princeton.edu/~cmi/research.

241 "Enoch Durbin": Letter from Enoch Durbin to the author, used with permission.

242 "Despite the jibes": See Peter Hoffmann's *Tomorrow's Energy*.

242 "As for": The Bain story is found in the Hoffmann book.

243 "Nelly Rodriguez": See papers by Nelly Rodriguez and Terry Baker in *The Journal of Physical Chemistry*, February 15, 2001, and in other publications.

244 "America's": For more on the Department of Energy's hydrogen program, see www.eere.energy.gov/hydrogenandfuelcells/hydrogen.

246 "That is why": For a vision of how hydrogen and electricity could coexist in the energy future, see the Electric Power Research Institute's "Electricity Technology Roadmap," 1999.

247 "In the long term": See studies by America's National Renewable Energy Laboratory on producing hydrogen from renewable sources, the European Commission's "Hydrogen Vision" document, and Jeremy Rifkin's *The Hydrogen Economy*.

248 "The great": On sequestration, see the International Energy Agency's excellent report "Putting Carbon Back in the Ground."
248 "Some greens": The citation from *The Ecologist* is drawn from the issue of October 22, 2001. Robert Socolow's thoughts are found in the Aspen Institute book on climate change, *What Next?*
257 "But Rasul": *Red Herring*, July 2000.
258 "Nearly": Iain Carson, *The Economist*, October 31, 1998.

9: Rocket Science Saves the Oil Industry

Roger Anderson, Stephen Whittaker, and many kind people at Shell's operations in the Gulf of Mexico deserve thanks for help with this chapter.

264 "Higgins": See *The Spindletop Gusher*, available at the Gladys City museum in Texas.
266 "Even the discoveries": For example, see "The Great Game Gets Rough," *Washington Post*, January 26, 2000.
266 "To the disappointment": Good estimates of proven reserves can be found in the BP Statistical Review of World Energy, produced annually. Ironically, BP is a leading investor in the Caspian region and one of the companies bullish about its prospects.
267 "Veterans": *The Wall Street Journal* ran a story on July 3, 2000, describing how the oil industry's assumptions about the presence of oil under deep water have changed over the years.
268 "The International Energy Agency": The IEA's *World Energy Outlook 2001* forecasts oil demand and supply out to 2020.
271 "Researchers": For Schlumberger's view of the future of oil technology, see Euan Baird's comments at www.slb.com.
272 "Roger Anderson": See work by Roger Anderson at Columbia University's Lamont-Doherty Earth Observatory for more details on how the oil industry can learn from the experience of Boeing in virtual modeling (www.ldeo.columbia.edu).

10: A Renaissance for Nuclear Power?

Walt Patterson and Malcolm Grimston were helpful in preparing this chapter. Much of the analysis of nuclear economics was drawn from Double or Quits *and* Nuclear Power in the OECD.

275 "Richard Thornburgh": Richard Thornburgh, interview on the website of the *Washington Post*: http://discuss.washingtonpost.com/wpsrv/zforum/99/thornburgh0329.htm.
277 "Some politicians": Loyola de Palacio, interview in *The Business Link*, Spring 2002.
277 "In fact": See "The Role of Nuclear Power in Western Europe," by

Florentin Krause and Jonathan Koomey, in *Energy Policy in the Greenhouse* (International Project for Sustainable Energy Paths, 1994).

282 "In one": A full account of the local revolt is found on LIHistory.com.

288 "Throughout the world": See studies on the decline of nuclear engineering programs at universities done by the OECD's Nuclear Energy Agency (www.nea.fr) and *IEEE Spectrum* (www.spectrum.ieee.org).

11: Micropower Meets Village Power

Special thanks go to Christine Eibs Singer, Navroz Dubash, Jason Edwin Adkins, and Charles Feinstein for help with this chapter. A good introduction to the topic of energy poverty is found in the World Energy Assessment.

292 "At a gathering": Bill Gates spoke at the Digital Dividends Conference in Seattle, Washington, October 2000.

292 "Bridging": The Village Power conference referred to ("Empowering People and Transforming Markets") was held at the World Bank's headquarters in Washington, D.C., in December 2000.

293 "One way": From the *World Energy Assessment.*

294 "The International Energy Agency": The IEA's study on energy poverty was released in August 2002, just before the Johannesburg Earth Summit; the analysis also made its way into the *World Energy Outlook* released later that year.

295 "Energy consumption": Energy growth rates to 2020 are found in the IEA's *World Energy Outlook 2001.*

296 "Some officials": On China's hydrogen efforts, see work at Tsinghua University in Beijing.

297 "We are at": See the final report from the G-8 Renewable Energy Task Force, (www.renewabletaskforce.org), led by Sir Mark Moody-Stuart, former chairman of Royal Dutch/Shell.

297 "Some groups": See "The Climate of Export Credit Agencies," by Crescencia Maurer with Ruchi Bhandari, published by the World Resources Institute in May 2000.

301 "That report's author": Conversation with author of IEA book *Electricity in India*, Paris, 2002.

301 "Navroz Dubash": See "Power Politics" by World Resources Institute (www.wri.org).

304 "Dam Shame": See Arundhati Roy's book *The Cost of Living.*

306 "That commission's final report": See *Dams and Development.*

307 "Less aid": For more on the NGO perspective on big dams, see work by the International Rivers Network (www.irn.org).

308 "A reliable study": See "Looking at Energy Subsidies: Getting Prices Right," IEA, 1999.

310 "The rallying cry": Frank Tugwell's call to arms was uttered at the Village Power conference in 2000.

310 "India's Centre": "Breaching the Digital Divide: Can Indian Villages Be Logged On to the Infotech Highway?" *Down To Earth*, published by India's Centre for Science and Environment (www.cseindia.org).

311 "In San Ramón": The claims were made by Gerardo Zepeda at the Village Power conference in 2000.

312 "Families in Peru's": Figures rely on analysis by E+Co.

312 "In Yemen": The World Bank has examined the curious case of Yemen's power market in "Free Entry in Infrastructure" (Working Paper 2093, March 1999).

Epilogue: The Future's a Gas

Thanks to Klaus Lackner and Joan Ogden for help with this chapter.

317 "As far back": Klaus Lackner of Columbia University argues that the Verne quote, while evocative, is not quite right in technical terms: "This part of the quote is the source of much confusion. Water is not the coal of the future, it is just the opposite. Water is to hydrogen as CO_2 is to coal."

318 "Hydrogen can": BMW, for example, is betting heavily on modified internal combustion engines, not fuel cells, as the hydrogen power plants of the future.

318 "Jesse Ausubel": Jesse Ausubel's article on decarbonization appeared in the Summer 1996 issue of *Daedalus* (www.daedalus.amacad.org/issues/su96rel.html).

318 "It is encouraging": Quote drawn from Robert Hefner's *The Age of Energy Gases.*

319 "The International Energy Agency": The IEA reference is from *World Energy Outlook 2001.*

319 "Royal Dutch/Shell": The Shell Scenarios to 2050 were published in 2001.

324 "One recent": The encounter on Capitol Hill took place the summer of 2001 during a hearing on the Corporate Average Fuel Economy standards.

324 "Making a pointed reference": To see why environmental externalities justify higher gasoline taxes in the United States (and perhaps lower taxes in Britain), peruse the economic analyses done by Resources for the Future and by the Energy Institute at the University of California on this topic.

BIBLIOGRAPHY

Adelman, Morris. "World Oil: The Clumsy Cartel." *The Energy Journal*. London: Energy Economics Education Foundation, 1980.

Barnett, Harold J., and Chandler Morse. *Scarcity and Growth: The Economics of Natural Resource Availability*. Baltimore: Johns Hopkins University Press, 1963.

Barrett, Scott. *Environment and Statecraft: The Strategy of Environmental Treaty-Making*. New York: Oxford University Press, 2003.

Beck, Peter, and Malcolm C. Grimston. *Double or Quits? The Global Future of Civil Nuclear Energy*. London: Royal Institute of International Affairs, 2002.

Bossel, Ulf. *The Birth of the Fuel Cell, 1835–1845*. Oberrohrdorf, Switzerland: European Fuel Cell Forum, 2000.

Bredeson, Carmen. *The Spindeltop Gusher: The Story of the Texas Oil Boom*. Brookfield, Conn.: Millbrook Press, 1996.

British Petroleum. *BP Statistical Review of World Energy*. London: BP, published annually.

Casten, Thomas, and Sean Casten. "Transforming Electricity." *Northeast-Midwest Economic Review*, November/December 2001.

Cheney, Margaret. *Tesla: Man out of Time*. New York: Simon and Schuster, 2001.

Chertow, Marian, and Daniel Esty. *Thinking Ecologically: The Next Genera-*

tion of Environmental Policy. New Haven: Yale University Press, 1997.

Crew, Michael, and Joseph Schuh, eds. *Markets, Pricing and Deregulation of Utilities*. Boston: Kluwer Academic Publishers, 2003.

Daily, Gretchen C., and Katherine Ellison. *The New Economy of Nature*. Washington, D.C.: Island Press/Shearwater Books, 2003.

Dasgupta, Partha. "Population, Poverty and the Natural Environment." In *The Handbook of Environmental and Resource Economics*, edited by Jeroen C.J.M. Van Den Bergh. Amsterdam: North Holland-Elsevier Science, 2003.

Davis, Devra. *When Smoke Ran Like Water*. New York: Basic Books, 2002.

Dunn, Seth. "Hydrogen Futures: Toward a Sustainable Energy System." *Worldwatch Paper 157*. Washington, D.C.: Worldwatch Institute, 2001.

———. "Micropower: The Next Electrical Era." *Worldwatch Paper 151*. Washington, D.C.: Worldwatch Institute, 2000.

Easterbrook, Gregg. *A Moment on the Earth: The Coming Age of Environmental Optimism*. New York: Viking Penguin, 1996.

Esty, Daniel, and Peter Cornelius, eds. *Environmental Performance Measurement: The Global Report 2001–2002*. New York: Oxford University Press, 2002.

"Exploring the Future: Energy Needs, Choices, and Possibilities, Scenarios to 2050." Report, Royal Dutch/Shell Group, 2001.

Fox, Loren. *Enron: Rise and Fall*. Hoboken, N.J.: John Wiley and Sons, 2003.

Gayoom, Maumoon Abdul. *The Maldives: A Nation in Peril*. Republic of Maldives: Ministry of Planning, 1998.

Grubb, Michael, Christiaan Vrolijk, and Duncan Brack. *The Kyoto Protocol: A Guide and Assessment*. London: Earthscan Publications, 1999.

Hawken, Paul, Amory B. Lovins, and Hunter L. Lovins. *Natural Capitalism: Creating the Next Industrial Revolution*. Boston: Little, Brown, 1999.

Heal, Geoffrey. *Nature and the Marketplace*. Washington, D.C.: Island Press, 2000.

Hefner, Robert A., III. *Energy and the U.S. Marketplace: Toward Environmentally Sustainable Economic Growth*. Oklahoma City: GHK Company, 2000.

Hertsgaard, Mark. *Earth Odyssey: Around the World in Search of Our Environmental Future*. New York: Broadway Books, 1999.

Hoffmann, Peter. *Tomorrow's Energy: Hydrogen, Fuel Cells and the Prospects for a Cleaner Planet*. Cambridge, Mass.: MIT Press, 2001.

Howarth, Stephen. *A Century in Oil: The "Shell" Transport and Trading Company 1897–1987*. London: Weidenfeld and Nicolson, 1997.

International Energy Agency. *Competition in Electricity Markets*. Paris: IEA, 2001.

———. *Distributed Generation in Liberalised Electricity Markets*. Paris: IEA, 2002.

———. *Electricity Market Reform*. Paris: IEA, 1999.

———. *Nuclear Power in the OECD*. Paris: IEA, 2001.

———. *Toward a Sustainable Energy Future*. Paris: IEA, 2001.

———. *World Energy Outlook*. Paris: IEA, published annually.

Khoshoo, T. N. *Mahatma Gandhi: An Apostle of Applied Human Ecology*. New Delhi: TATA Energy Research Institute, 1995.

Lomborg, Bjorn. *The Skeptical Environmentalist: Measuring the Real State of the World*. Cambridge, Eng.: Cambridge University Press, 2001.

Lovins, Amory B., et al. *Small Is Profitable: The Hidden Economic Benefits of Making Electrical Resources the Right Size*. Snowmass, Colo.: Rocky Mountain Institute, 2002.

McNeill, J. R. *Something New Under the Sun: An Environmental History of the Twentieth-Century World*. New York: W. W. Norton, 2000.

McPhee, John. *Encounters with the Archdruid*. New York: Farrar, Straus and Giroux, 1971.

Molina, Luisa, and Mario Molina, eds. *Air Quality in the Mexico Megacity: An Integrated Assessment*. Boston: Kluwer Academic Publishers, 2002.

Moore, Stephen, and Julian Simon. *It's Getting Better All the Time: The 100 Greatest Trends of the Last 100 Years*. Washington, D.C.: Cato Institute, 2000.

Myers, Norman. *Perverse Subsidies: Tax $s Undercutting Our Economies and Environments Alike*. Winnipeg, Can.: International Institute for Sustainable Development, 1998.

Organization for Economic Cooperation and Development. *Improving the Environment Through Reducing Subsidies*. Paris: OECD, 1998.

———. *Innovation and the Environment*. Paris: OECD, 2000.

———. *OECD Environmental Outlook*. Paris: OECD, 2001.

———. *Sustainable Development—Critical Issues*. Paris: OECD, 2001.

Patterson, Walt. *Transforming Electricity: The Coming Generation of Change*. London: Earthscan Publications, 1999.

Pew Center on Global Climate Change. *Climate Change: Science, Strategies, and Solutions*. Leiden, Netherlands: Brill Academic Press, 2001.

Probst, Katherine, and David Konisky. *Superfund's Future*. Washington, D.C.: Resources for the Future, 2001.

Rifkin, Jeremy. *The Hydrogen Economy: The Creation of the Worldwide Energy Web and the Redistribution of Power on Earth*. Cambridge, Eng.: Polity, 2002.

Riggs, John A., ed. *U.S. Policy on Climate Change: What Next?* Queenstown, Md.: The Aspen Institute, 2002.

Roy, Arundhati. *The Cost of Living?* New York: Modern Library, 1999.

Shabecoff, Philip. *Earth Rising: American Environmentalism in the 21st Century*. Washington, D.C.: Island Press, 2000.

Smithsonian Institution. *Powering a Generation of Change*. Internet-based project on America's power deregulation, http://americanhistory.si.edu/csr/powering.

South Coast Air Quality Management District. *The Southland's War on Smog*. Los Angeles: SCAQMD, 1997.

Stavins, Robert N., ed. *Economics of the Environment: Selected Readings*. New York: W. W. Norton, 2000.

Stevens, William K. *The Change in the Weather: People, Weather, and the Science of Climate*. New York: Delacorte Press, 1999.

United Nations Department of Economic and Social Affairs, and World Energy Council, *World Energy Assessment*. New York: UNDP, 2000.

United Nations Intergovernmental Panel on Climate Change. *Climate Change 2001: Third Assessment Report*. Cambridge, Eng.: Cambridge University Press, 2001.

Verleger, Philip, Jr. *Adjusting to Volatile Energy Prices*. Washington, D.C.: Institute for International Economics, 1993.

Victor, David G. *The Collapse of the Kyoto Protocol and the Struggle to Slow Global Warming*. Princeton, N.J.: Council on Foreign Relations/Princeton University Press, 2001.

Walsh, James. *The $10 Billion Jolt: California's Energy Crisis: Cowardice, Greed, Stupidity and the Death of Deregulation*. Los Angeles: Silver Lake Publishing, 2002.

World Bank. *China: Air, Land and Water—Environmental Priorities for a New Millennium*. Washington, D.C.: World Bank, 2001.

———. *Greening Industry: New Roles for Communities, Markets, and Governments*. Washington, D.C.: World Bank, 1999.

——— . *Health and Environment*. Washington, D.C.: World Bank, 2001.

———. *World Development Report 2003*. Washington, D.C.: World Bank, 2002.

World Commission on Dams. *Dams and Development: A New Framework for Decision-Making*. London: Earthscan Publications, 2000.

Yergin, Daniel. *The Prize: The Epic Quest for Oil, Money and Power*. New York: Simon and Schuster, 1991.

Yergin, Daniel, and Joseph Stanislaw. *The Commanding Heights: The Battle for the World Economy*. New York: Simon and Schuster, 2002.

ACKNOWLEDGMENTS

I simply could not have produced this book without the ideas and inspiration drawn from my colleagues, teachers, friends, and family. If I have been able to see beyond my nose on these pages, it is only because I am stepping on the toes of giants.

First, I thank my colleagues at *The Economist*. I am extremely grateful to Bill Emmott, the editor-in-chief, for giving me the time to write *Power to the People* and for giving me permission to use articles that I have written for the magazine. I thank John Peet for tolerating my absences and for his keen insights on the book. Various gifted editors have helped me to refine my thinking on energy matters over the years; chief among them are Edward Carr, Matthew Valencia, and Barbara Beck. Daniel Franklin and John Micklethwait offered advice and encouragement. Zanny Minton Beddoes, Nick Valery, Natasha Loder, and John Smutniak were kind enough to read various parts of the book, and to offer helpful suggestions.

A great many people outside *The Economist* have also taken the trouble to educate me about the energy world. In particular, I thank Amory Lovins, Walt Patterson, and Paul Portney for sharing with me the insights that they have gleaned over a lifetime. A number of experts are cited directly in the text of this book. Others offered comments on parts of the book, and their names are found in the relevant chapter of the Notes. Yet others have asked to remain anonymous. I am grateful to each and every one of these generous people.

I have been exceptionally lucky to have had a first-rate professional team working on this book. My agent, Dan Mandel, has been a source of clearheaded advice and steady encouragement. Ethan Nosowsky, my patient and perceptive editor at Farrar, Straus and Giroux, has done much to breathe life into this work. Thanks go also to the fine team of copy editors, graphic designers, marketing experts, and others at FSG who have worked on *Power to the People*. Michael Coulman, a veteran researcher at *The Economist*, spent countless late nights checking facts and tracking down obscure sources. For his hard work and his ingenuity, I thank him profusely. Sam Skinner also helped me research parts of the book.

My family and friends will be grateful that this project is now over. They have so very kindly put up with my droning on about energy these past few years. I am especially grateful to my parents, Boothipuram and Chitra Vaitheeswaran, and my sister, Shoba, for their unflagging devotion. The memory of my mother, Seetha, and of her faith in me served as an inspiration in my darkest hours. I thank Kavitha Prakash for her early support for this project. A number of friends, most notably Dan Dombey and Rana Sarkar, suffered through impenetrable drafts of this work and provided useful comments. Dan also offered these early words of encouragement: "Remember that the only thing more boring than the power industry is the micropower industry." Readers who find that this book has strayed too far toward the picaresque now know whom to blame.

Finally, I would like to thank Claudia Kolker and Iain Carson. It was Claudia who persuaded me that there was indeed a book knocking around in my head that needed to be written. She introduced me to Dan Mandel, my agent. And she proved a fount of encouragement and advice that ensured that this book actually saw the light of day. No journalist could ask for a more magnanimous mentor than Iain, the industry editor of *The Economist*. Over the years, he has been a sounding board for story ideas, an advocate of my cause, and a dear friend. What is more, he was also the first person ever to utter these two magical words to me: *fuel cells*. Without Claudia and Iain, there would have been no book.

To them, to all my family, I dedicate *Power to the People*.

INDEX